Exploring Biology in the Lab

Edited by Rita Connolly
and the Biology Department
of Camden County College

KENDALL/HUNT PUBLISHING COMPANY
4050 Westmark Drive Dubuque, Iowa 52002

Cover image credits:
Background image © Getty Images
Beakers, amoebas, and windmills © JupiterImages Corporation

Copyright © 2006 by Camden County College Biology Department

ISBN 13: 978-0-7575-3553-6
ISBN 10: 0-7575-3553-4

Kendall/Hunt Publishing Company has the exclusive rights to reproduce this work,
to prepare derivative works from this work, to publicly distribute this work,
to publicly perform this work and to publicly display this work.

All rights reserved. No part of this publication may be reproduced,
stored in a retrieval system, or transmitted, in any form or by any
means, electronic, mechanical, photocopying, recording, or otherwise,
without the prior written permission of the copyright owner.

Printed in the United States of America
10 9 8 7 6 5 4 3 2 1

Contents

Introduction v
Laboratory Safety Agreement xi
Laboratory Drawings xiii
Experimental Laboratory Reports xv

LAB EXERCISES

I Scientific Method

Exercise 1 Methods of Science	3
Exercise 2, Part 1: Introduction to the Information Literacy Project	9
Part 2: Information Literacy—Abstract Assignment	11
Part 3: Information Literacy—Writing Assignment	13
Exercise 3 Scientific Vocabulary	15

II Chemistry

Exercise 4 Laboratory Measurements	21
Exercise 5 Measurement	29
Exercise 6 Enzyme Activity	33
Exercise 7 Food Nutrient Analysis	37

III Cells

Exercise 8 Using the Microscope	43
Exercise 9 Microscopic Techniques	49
Exercise 10 Plant Cells	55
Exercise 11 Animal Cells	59
Exercise 12 Diffusion: Osmosis and Dialysis	63

IV Metabolism

Exercise 13, Part 1: Photosynthesis	73
Part 2: Photosynthesis Laboratory Report	77
Exercise 14 Separation of Leaf Pigments by Paper Chromatography	79
Exercise 15 Fermentation Lab	83
Exercise 16 Cell Respiration— Go for the Burn!!	87

V Cell Division

Exercise 17 Cell Division	91

VI Genetics

Exercise 18, Part 1: Student Worksheet DNA: The Genetic Code	103
Part 2: Student Worksheet RNA: The Code Transcribed	107
Part 3: Protein Synthesis: The Code Translated	111
Exercise 19 Macromolecular Model Lab	113
Exercise 20 Genetics	115
Exercise 21 Human Chromosome Analysis BioKit®	117
Exercise 22 ABO-Rh Blood Typing	123
Exercise 23 Human Genetic Traits	131
Exercise 24 Genetic Problems	135
Set 1	135
Set 2	137
Set 3	141

VII Evolution

Exercise 25 A Simulation of Natural Selection	145

VIII Survey of Kingdoms

Exercise 26 Observation of Prokaryotic Cells	151
Exercise 27 A Simulated Epidemic	155

Exercise 28, Part 1: Kingdom Protista—
    Algae   157
        Part 2: Kingdom Protista—
        Protozoa and Slime
        Molds   167
Exercise 29 Kingdom Fungi   175
Exercise 30 Cultivation
    of Mushrooms   189
Exercise 31 Kingdom Plantae   191
Exercise 32 Botanical Field Project   201
Exercise 33 Leaf and Tree
    Identification   209

Appendices
Periodic Table   211
Laboratory Orientation   213
Molecular Genetics   217

Introduction

This laboratory manual was prepared by the faculty and staff of the Biology Department at Camden County College to assist the student in the study of Biology. The department would like to acknowledge and thank Mrs. Patricia Farley, the department secretary and Mrs. Stacie Cantu, the Biology technician for their help in assembling and creating material for this manual.

HOW TO STUDY FOR BIOLOGY 111

Biology 111 is a course designed to teach the fundamentals of Biology to science majors. It is a fast paced, intensive course that can be challenging for the best of students. The following suggestions may help you in your studies:

1. Attend class regularly. There is no substitute for in person attendance. Many times, a missed assignment, quiz, or lab is something that cannot be made up. Poor attendance will affect your grade.
2. Read the course outline that your instructor gave you at the start of the semester. It contains important information on attendance, due dates, and testing policies.
3. Be prepared before you come to class. You should make every attempt to read the assigned material BEFORE your lecture.
4. Take notes in class—even if your instructor has provided a lecture outline or set of Power Point notes. Your own notes are crucial to understanding the material.
5. Read the lab exercises BEFORE you come to class. This will help you to perform the lab more efficiently. It will also allow you to ask questions before the lab if you are confused about the exercises.
6. Do not procrastinate. BIO-111 is a content heavy course. The best way to study is to review your notes in small segments as frequently as possible. You cannot wait until the last minute to study for an exam. The concepts in Biology do not lend themselves to cramming.
7. If you have problems understanding the work, ask for help. Your instructor will explain the material in class, but you should try to ask directed, concrete questions.
8. There is FREE TUTORING available at the College Tutoring Center.
9. Get to know your lab partners and other classmates. Obtain the telephone numbers or e-mails of at least three classmates so you can keep up with class if you are absent, or need to clarify an assignment, etc. It is also helpful to form study groups.
10. Devote enough time to learn the material. The suggestion for beginning science classes is: 3 hours outside the class for every hour inside the class. This means that you should expect to spend about 15–18 hours/week on this class.

Biology Department Laboratory Safety Procedures for Biology Courses

I. SAFETY PROCEDURES FOR ALL BIOLOGY LABORATORIES

The risks incurred in the biology laboratories of Taft Hall (T301, T303, T304), Truman (Tr122, Tr125), Helene Fuld (HF210), and Camden 329 are minimal and usually associated with the occasional use of open flames, sharp objects, some noxious chemicals, and attenuated or non-virulent strains of bacteria. Safety precautions are listed below and must be adhered to in all biology laboratories.

Each student and new instructor must receive a copy of the Biology Laboratory Safety Procedures at the beginning of each semester, read the policy, and sign the Laboratory Safety Agreement. Instructors will submit all completed forms to the biology office.

A. General

1. Work carefully and cautiously in the laboratory, using common sense and good judgment at all times.
2. EATING, DRINKING, AND SMOKING ARE PROHIBITED in the laboratory and in the laboratory space of a combined lecture/laboratory room.
3. Long hair must be tied back during laboratory sessions.
4. Footwear must completely cover the foot. Shoes with open toes, open heels, or sandals are prohibited.
5. No sleeveless tops are permitted. Thighs and midriffs must be covered with protective clothing while working in the laboratory. Lab coats must be worn when directed by the instructor.
6. Identify the location of all exits from the laboratory and from the building.
7. Be familiar with the location and proper use of fire extinguishers, fire blankets, first aid kits, spill response kits, and eye wash stations in each laboratory.
8. Note the location of the red phones that provide direct access to the Office of Public Safety. In the event of an emergency, pick up the red receiver and state the location and the nature of the emergency. The Office of Public Safety can also be reached by dialing extension 7777. Identify the location of the nearest desk phones.
9. Report all injuries, spills, breakage of glass or other items, unsafe conditions, and accidents of any kind, no matter how minor, to the instructor immediately.
10. Keep sinks free of paper or any debris that could interfere with drainage.
11. Lab tables must be clear of all items that are not necessary for the lab exercise.
12. Wash hands and the lab tables with the appropriate cleaning agents before and after every laboratory session.

B. Open Flames—Fire Hazard

1. Identify and be familiar with the use of dry chemical fire extinguishers that are located in the hallways and laboratory rooms.
2. Flames are only to be used under the supervision of the instructor.

C. Sharp Objects and Broken Glass

1. Pointed dissection probes, scalpels, razor blades, scissors, and microtome knives must be used with great care, and placed in a safe position when not in use.
2. Containers designated for the disposal of sharps (scalpel blades, razor blades, needles; dissection pins, etc.) and containers designated for broken glass are present in each laboratory. Never dispose of any sharp object in the regular trash containers.
3. Report all cuts, no matter how minor, to the instructor. Students with moderate to extensive bleeding will be referred to the Office of Public Safety (red phone or extension 7777).
4. All biology labs and the biology preparation room (T305) house a first aid kit containing antiseptics, bandages, Band-Aids, and gloves to care for minor cuts.
5. Do not touch broken glass with bare hands. Put on gloves and use a broom and dustpan to clean up glass. Dispose of ALL broken glass in the specific container marked for glass. Do not place broken glass in the regular trash.
6. When cutting with a scalpel or other sharp instrument, forceps may be used to help hold the specimen. Never use fingers to hold a part of the specimen while cutting.
7. Scalpels and other sharp instruments are only to be used to make cuts in the specimen, never as a probe or a pointer.

D. Noxious Chemicals

1. Material Safety Data Sheets (MSDS) are available in a yellow binder mounted on the door of the biology prep room (T305) in Blackwood. In Camden, the MSDS is in a marked filing cabinet in the office area of 329. In case of a spill, an accident, or a safety question, students and instructors can find chemical safety information in the MSDS.
2. The biology prep room in Taft is equipped with a portable safety exhaust hood for the handling of noxious fumes. Room 329 in Camden is equipped with three safety hoods.
3. Chemical spill cleanup kits are available in every biology lab and biology prep room.

E. Instruments and Equipment

Care must be used when handling any equipment in the laboratory. Students are responsible for being familiar with and following correct safety practices for all instruments and equipment used in the laboratory.

MICROSCOPE HANDLING

1. Microscopes must be carried upright, with one hand supporting the arm of the microscope and the other hand supporting the base. Nothing else should be carried at the same time.
2. Microscope must be positioned safely on the table, NOT near the edge.
3. After plugging the microscope into the electrical outlet, the cord should be draped carefully up onto the table and never allowed to dangle dangerously to the floor.
4. The coarse adjustment must NEVER be used to focus a specimen when the 40X or oil immersion lens is in place.
5. When finished with the microscope, the cord should be carefully wrapped around the microscope before returning it to the cabinet.
6. The microscope must be placed upright and in the appropriate numbered slot in the cabinet.
7. All prepared microscope glass slides are to be returned to their appropriate slide trays; wet mount preparations are to be disposed of properly.

8. Malfunctioning microscopes should be reported to the instructor.

HOT PLATES AND WATER BATHS
1. The instructor will regulate the temperature of hot plates and water baths with a thermometer.
2. This equipment must be placed in a safe place.
3. Use insulated gloves or tongs to move beakers or test tubes in and out of the water baths.
4. Use care when working near hot plates and water baths, as they may still be hot even after being turned off.

F. Preserved Specimens
1. Gloves (latex and nonlatex) are provided to handle preserved specimens.
2. When larger specimens are being dissected, the part of the specimen that is not being dissected should be kept enclosed in the plastic bag.
3. When dissecting smaller specimens, seal the bag after removing the specimen, so as to confine the preservative in the specimen bag.
4. Notify the instructor if there is a spill of preservative.
5. Body parts or scraps of the specimen are NOT to be disposed of in the sink.
6. Dispose of dissecting pins or other sharp objects in the red sharps containers, NOT in the regular trash.
7. Specimens are to be clearly labeled and stored in designated containers or cabinets when not in use.
8. Follow the directions of the instructor concerning the proper disposal of preserved specimens after they are finished being used.

G. Body Fluids
Special precautions are to be followed in all laboratories using any body fluids, such as blood, saliva, and urine, because of the potential to transmit disease-causing organisms.

1. Follow all instructions carefully.
2. Use gloves and goggles in all laboratory experiments that involve the use of body fluids.
3. All contaminated material, such as slides, coverslips, toothpicks, lancets, alcohol swabs, etc., must be placed in a biohazard bag for proper disposal and should never be reused.
4. No animal tissues or samples of body fluids are to be brought into the laboratory from outside sources.

II. BIOLOGY FIELD TRIP SAFETY PROCEDURES
The following laboratory safety guidelines for Biology field trips are in addition to the laboratory safety procedures to be followed for all biology laboratories:

1. **Appropriate dress**
   a. Wear clothing that may potentially get dusty, muddy, or wet.
   b. Wear closed shoes, such as sneakers or hiking boots.
   c. Depending on conditions, a rain poncho may be a good idea.
2. Prior to the program, inform the instructor of any known environmental allergies. This includes (but is not limited to) poison ivy, poison sumac, bee/wasp stings, or ant bites.

3. If you have asthma, or other disorders that may be environmentally stimulated, make certain to bring along the prescribed inhaler, hypodermic solution, or other forms of medical treatment.
4. **Take precautions against ingesting any unknown plant material:**
   a. Do not eat unknown or unidentified plant materials such as berries, seeds, fruits, mushrooms of any type, sap, or resins.
   b. Do not rub unknown plant materials or juices on your skin.
   c. Do not inhale or expose your skin or eyes to the smoke of a burning plant.
   d. Do not eat food after handling plants until you've washed your hands.

Note: The 24-hour emergency phone number for the ***NJ Poison Control Center*** is

(800) 222-1222

This number can be dialed from a cell phone in the field.

The Camden County Dept. of Public Health has posted a *Health Alert,* noting various tips on prevention of and dealing with plant poisoning. The *Health Alert* can be found at:

http://www.co.camden.nj.us/health/alerts_plants.html

5. **Animal pests**
   a. Depending on conditions on the day of the trip, you might consider using an insect repellant.
   b. Check for ectoparasites (external parasites), especially ticks, on both skin and clothing. They prefer warm, hidden regions, such as the scalp or armpits.
   c. Once you've returned home, it is advisable to check more thoroughly.

6. **Specimen collection and environmental preservation**

   Do not handle or provoke wild animals.

   - Be conservative in collecting specimens for the course.
   - Take limited amounts of the specimens being studied, and no others.
   - Remain on the trails, unless asked to move elsewhere.
   - If a trail comes to a fork or cross-way, make certain the person behind you knows the correct direction.

7. Report all injuries to the instructor.

LABORATORY SAFETY AGREEMENT

I am enrolled in: *(Course name)* _____

(Course number) _____

My instructor's name is: _____

I, _____, have carefully read and understand the biology science laboratory safety procedures of Camden County College. I agree to adhere to these guidelines, and realize that it is my responsibility to do so, for my own safety and the safety of all others.

Print name: _____ *Date:* _____

Signature: _____

Laboratory Safety Procedures must be distributed during the first day of class to each student. The Laboratory Safety Agreement must be signed and returned to the instructor before the start of the first lab. It is the instructor's responsibility to make sure that each student in the class has signed the laboratory safety agreement. The instructor is responsible for turning in the signed form for each student in each class to the biology office.

Laboratory Drawings

1. It is essential that laboratory drawings be maintained. Your laboratory drawings serve as a permanent record of your laboratory observations. Laboratory drawings will be collected at the end of each laboratory period and will be returned at the next period. Attendance will be taken by this means.
2. Laboratory Drawings
    a. All laboratory drawings must be made on white, unlined, heavy bond paper 8 1/2 × 11 inches.
    b. All drawings should be done with a hard lead (3H or 4H) pencil. Drawings done with soft pencil or in ink will smudge and will not be acceptable.
    c. All drawings must be single line, outline drawings. Shading or cross hatching is not permitted. Solid areas may be done by stippling.
    DO NOT USE COLORED PENCILS, INK, OR CRAYONS.
    d. Each drawing must be labeled with the following legends:

    Common name of specimen

    Phylum, class, and genus

    Magnification of drawing (if applicable)

    Location or orientation of view drawn

    In addition, each individual drawing must have the important structures labeled with identification lines (leaders) running to the structure to be labeled. Leader lines should not cross each other.
    e. Each sheet of drawings must have the following headings at the top of the page:

Left Hand Top	**Right Hand Top**
Lab exercise title and lab exercise	Student's name
Lab exercise number	Date of drawing
Laboratory section	

    f. All labels, legends, titles, and headings must be printed.
3. All drawings must conform to these criteria or with a suitable format approved by the laboratory instructor. Failure to comply will result in the drawing to be graded as unacceptable.

Experimental Laboratory Reports

Reports for experimental labs should be written in the style of a research journal article. The report should contain an appropriate title and separate sections as follows:

Purpose

Materials

Methods

Results or Findings

Discussion

References (if applicable)

Purpose—The purpose should be short and describe the intent.

Materials should be written in a list format. Include all materials that you used for the lab, including glassware, reagents, specimens, etc.

Methods should describe exactly what was done, including what controls were used.

Results should only describe the data, no discussions or explanations. If you recorded results on a chart or on a graph, this is the section where you would display those results.

Discussion—The discussion section is where the data or findings are explained. Ask yourself: What did the experiment show? Were the results as expected? If not, why not? Are further investigations indicated? What can you conclude from your experiments? Don't make assumptions. Be sure that your conclusions really follow from your data.

Lab reports must be typed, double-spaced, and stapled. Due dates will be discussed in class. Although most experimental labs will be done in groups, each individual must write his or her own lab report independently of the others in the group. Plagiarism on lab reports may result in a loss of points, or you may receive a zero for the lab. Your instructor will discuss the penalties for plagiarism that apply to your class.

Part I

Scientific Method

Name _____ Section _____ Date _____

Exercise 1: Methods of Science

OBJECTIVES

After completing this exercise, you should be able to:

1. List and define the steps of the scientific method.
2. State the purpose of an experiment.
3. Differentiate between experimental and control groups of an experiment.
4. Using a series of preliminary observations:
   a. state a problem developed from these observations,
   b. formulate a hypothesis or probable solution to the problem,
   c. design an experiment to test the hypothesis.

INTRODUCTION

We live in a society that has changed drastically in the last 100 years. Many of these changes have been brought about by advances in science. Advances like the A-bomb, transistors, synthetic clothes, and thousands of other items were developed in physics and chemistry. In the next 50–100 years the changes that will probably most drastically effect everyone's personal lives will be advancements in biology. These changes have already begun to take place with organ transplants, body control through biofeedback, environmental control, genetic engineering, cloned sheep, cloned cows, cloned pigs, and thousands of other areas.

A basic understanding of how science, specifically biology, functions will be a valuable assistance to citizens. Knowing the anatomy of a fetal pig enables us to understand the similarities between humans and other vertebrates. Knowing the evolutionary relationship of humans to other animals can allow a better understanding of how people fit into the environment. An understanding of how DNA operates would be useful in trying to decide if you want to support governmental control over genetic experiments. Knowing that ultraviolet light causes mutations would be helpful in deciding if controls should be placed on nitrogen oxide emissions from automobiles that destroy the ozone layer and increase our UV exposure.

These problems and many others will be faced in the future. Can there be such a thing as too many people? At what point does this happen and what causes it? Once causes are established, be it for overpopulation or environmental deterioration, can probable solutions be devised and implemented? Answers to some of these questions will require a trained biologist. However answers to most of these questions and other problems will depend on the personal attitudes and understanding of everyone. Therefore, knowing how science functions is of importance. But more important is the realization that the technique a scientist employs in solving a problem is not limited to science. Many times this scientific technique or method is applied elsewhere.

From *Biological Investigations: Lab Exercises for General Biology* 11th edition by Edward Devine. Copyright © 2004 by Kendall/Hunt Publishing Company. Reprinted by permission.

PART I SCIENTIFIC METHOD

Progress in science comes as a result of scientific investigations. The methods used in these investigations are many and varied. The scientific method is a vivid, dynamic process and cannot be reduced to a simple formula. The classification system used here is only a broad general view that is helpful in orientating a beginning scientist.

1. **Observation** of a Phenomenon.

    Something is noticed and attention is given to the observation. In this experiment, you will be given a sealed box. After an initial observation you will notice that there is something inside the box.

2. **Problem or Question**

    A problem is defined or **a question is asked about the observation.** Rather than simply accepting the observation and forgetting it, the questioning mind asks "why?" In this particular experiment, you might ask "what are the contents of the box?"

3. **Preliminary Information**

    In most types of scientific research, gathering **preliminary information involves a literature search** through books and journals. In this experiment, you gather preliminary information by tipping and shaking the box while listening to the sounds of the contents. As you listen you begin to get ideas about the contents of the box.

4. **Hypothesis**

    A hypothesis is a **possible solution** to the problem. It can also be explained as an **"educated guess."** After reviewing the preliminary information, a hypothesis is formed. The hypothesis must be stated in such a way that it is testable. To make a hypothesis statement like, "the contents of the box are pretty," is not adequate as it does not set up a basis of an experiment since it is not testable. A statement like "the box contains a round object" is more appropriate.

5. **Experiment**

    The primary purpose of the experiment is to **test** whether **the hypothesis** is correct. An experiment should test only one factor or variable while all other variables are kept constant. Every experiment must have a **control or standard.** A controlled experiment involves using two groups of the same kind of item or organism and treating them exactly the same except for the **variable** being tested. First test for the number of objects in the box. To get the most valid results there needs to be a comparison made to a constant standard. This is the control. In this case, the control would be an identical box with objects in it that you have set up. Conduct the experiment by comparing the sounds of the two boxes. Continue testing the contents of the box by considering such characteristics as number, size, shape, etc.

6. **Data**

    The results of data are collected from the experiment. Often the data are collected in a numerical form. **Numerical data can be recorded in a table and plotted in a graph. Data can also be collected by visual and auditory means.** These data may be recorded in the written notes of the researcher.

7. **Discussion**

    The data are analyzed and their meanings are interpreted. Comparisons are often made with experiments and conclusions of other researchers.

8. **Conclusion**

    The conclusion **summarizes the results** of the experiment. The conclusion either accepts or rejects the hypothesis based on the data.

9. **Theory**

   A **theory** is a general concept that has been **proven true** by a large number of experiments and observations. There are many theories in biology, such as the theory of evolution, the cell theory, and the biogenesis theory. Theories can be modified as new information is added. For example, the theory of evolution proposed by Charles Darwin has been modified as knowledge of genetics has been gained.

PART II BLACK BOX EXPERIMENT

1. In this exercise you are presented with a sealed box. Take a few minutes to observe the box. You may handle the box and listen to it. You may not open the box. This exercise is commonly referred to as the "black box experiment." This box is meant to serve as a model for situations a scientist is likely to encounter. For example, scientists have been able to work out the structure of the atom without actually ever seeing an atom. If direct observation is not possible, alternate methods can be employed to collect information.

2. Your task is to determine the contents of the sealed box.

3. You are to set up an experiment and conduct it. You may do anything you wish to the box except open it. To assist you, there is a bag of assorted objects. Every object in your box is also in the bag plus lots of extra objects. Empty boxes are also provided. If there is anything you want to use to assist you, ask your instructor. If possible it will be provided.

   Caution: Boxes must never be untaped and opened.

   Gather the data of your observations and arrive at a conclusion. Record your conclusion about the contents of the box in the space provided.

   Conclusion: _____

   _____

   _____

   _____

   Circle the letter of your box: A, B, C, or D.

PART III SCIENCE ARTICLE

Read the following article. In the margin, identify the numbered portions of the article as:

   Observation
   Preliminary Information
   Hypothesis
   Experiment
   Control
   Variable
   Data
   Conclusion

COLD VIRUS KILLS CANCER

   1. *We have all noticed that a cold can strike the cells in the nose at any time.* **2.** *Now scientists have discovered methods to genetically engineer a common cold virus so it can strike brain tumor cells.* **3.** *Previous experiments have shown that for a cold virus to invade a cell, the virus must disable the cell's protective protein call Rb. Then the virus injects its own genetic material, which kills the cell by taking over the machinery of the cell.*

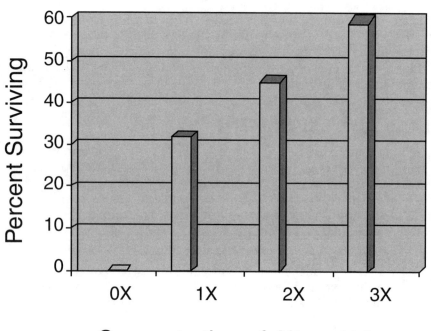

Figure 1.1
Survival Rate of Mice

4. Dr. Diana Wade of the Prairieville Research Group *altered the common cold virus by removing its Rb attack chemicals, and then injected the virus into mice with brain tumors.* The brain-tumor mice were separated into groups. **5.** One group was injected with the altered virus at a IX concentration. **6.** A second group received twice as much altered virus, a 2X concentration. **7.** A third group of mice was given a 3X concentration. **8.** The fourth group was injected with saline (a natural concentration of mild salt water). None of the groups showed signs of a cold. **9.** *Within 19 days, all of the saline-injected mice died, while up to 60% of the virus-injected mice remained alive and thriving* (**10.** *(Figure 1.1)*. To insure the surviving mice were tumor-free, the mice were euthanized and slides made of their brains. Empty cavities and scar tissue occupied spaces where tumors once grew.

11. Dr. Wade suspected the altered cold virus would only enter cancer cells because they have no Rb protective protein. **12.** Her research shows *there is a potential new tool for treating cancer.*

PART IV SCIENCE ARTICLE

Read the following article. In the margin, identify the numbered portions of the article as:

    Observation
    Preliminary Information
    Hypothesis
    Experiment
    Control
    Variable
    Data
    Conclusion

TAKE THAT, YOU DIRTY RAT

1. *Dr. John Vorva,* a neuropsychologist at the Chevy Institute, *has found that moderate stress can be good for immune system response.*

Stress is a well-known fact of life among high-pressure executives, working moms, and sleep-deprived college students. Besides making people ill-tempered, stress may also make people sick. **2.** *Many studies on animals and humans have shown that stress lowers immune system responses* by reducing the number of white blood cells and interleukin, their communication chemical.

Dr. Vorva suggests that stress many not be all bad. **3.** *Vorva exposed rats to different amounts of stress, then measured their white blood cell counts and interleukin levels.* **4.** Each day *the rats in one group were removed from their cages, petted, and placed in a small box for 10 minutes.* This was a mild variation from an otherwise rodent-like day of eating and playing. **5.** *The rats in the second group were petted, placed in the box where a grid on the floor delivered three mild electric shocks during the 10-minute experiment.* **6.** *The third group* was treated the same except they *received three long, strong, annoying electric shocks.* **7.** *The last group was left in their cages and not handled.*

8. Vorva suspected the increased amount of stress would result in increased immune suppression. At the end of the week-long experiment, white blood cell counts and interleukin levels were measured. **9.** As he expected, the *slightly stressed rats showed immunosuppression of about 10%.* **10.** The *highly stressed rats showed up to 70% immunosuppression.* **11.** *The moderately stressed rodents showed an amazing 30% increase in immune system response.*

12. Vorva's research shows *there is a range of stress that can be good for the immune system,* at least in rats. If this is true for humans, our modern way of life may not be as detrimental as many people think.

Name _____ Section _____ Date _____

Exercise 2, Part 1 — Introduction to the Information Literacy Project

PART I

As a department the Biology faculty are committed to help our students obtain a basic level of information literacy. The Middle States Commission on Higher Education in 2002 defined this as:

> . . . *an intellectual framework for identifying, finding, understanding, evaluating and using information. It includes determining the nature and extent of needed information; accessing information effectively and efficiently; evaluating information critically and its sources; incorporation of selected information in the learner's knowledge base and value system; using information effectively to accomplish a specific purpose; understanding the economic, legal and social issues surrounding the use of information and information technology; and observing laws, regulations, and institutional policies related to the access and use of information. (p. 32)*

We have partnered with the Library staff to address some of the goals established by the Middle States Commission and Camden County College regarding information literacy. This assignment must be completed by all Biology I students as professors build on this assignment in upper level courses. In order to communicate this to the faculty and our students we have prepared a list of goals, student learning outcomes, minimum requirements, and some approved assignment formats.

INFORMATION LITERACY GOALS, OBJECTIVES, AND ASSESSMENT

Goals
1. Incorporation of information literacy in Biology I
2. Develop research skills
3. Scientific literature acquisition and evaluation
4. To use scientific content to complete an assignment
5. Assessing skills and knowledge gained by students

Objectives/Student Learning Outcomes
1. Knowledge of the scientific method
2. Knowledge of the cycle of scientific literature
3. Ability to find and evaluate at least four pieces of scientific literature
4. Ability to use new information to complete a specific assignment
5. Completion of a bibliography in an approved format

Minimum Requirements of Assignment
1. Students will attend a presentation given by a librarian
2. Students will collect at least four pieces of scientific literature
3. Instructor will provide students with a guide detailing the assignment that they will assess
4. Instructor will provide grading criteria for the assignment
5. Students will be assessed by each instructor after completing one of the approved assignment formats below or by examination

Assignment Formats

1. Obtain four articles on one topic and write a mini-review. The articles may be divided into research and reviews as desired by the professor. The professor may request that some of the articles be obtained from the bound periodical collection at the Wolverton Resource Center or from the online databases. The student will summarize, not plagiarize, the articles and include a bibliography. We recommend that the student get the article preapproved by his/her instructor.

2. Obtain four articles on any Biology topics and abstract or summarize each article. The professor may wish to preapprove the articles as well as establish due dates throughout the semester.

3. The students are assigned groups and allowed to choose a topic or pick one from a list of approved topics. Each group researches and prepares an informed position on the topic and presents it to the class. Each group member may also be required to submit a written summary of the articles that they read.

4. The students prepare a mini study or précis where each student researches a specific hypothesis. The hypothesis must be simple and testable. After a literature review the student would suggest experiments to investigate their hypothesis.

Name _____ Section _____ Date _____

Exercise 2, Part 2: Information Literacy— Abstract Assignment

PART II
Scientific Publications

As you have learned, the scientific method is composed of a series of steps generally arranged in a distinct pattern; initial observation, hypothesis, further observation and where applicable, experiment and conclusion. There is yet one more step that is of prime importance. The capstone of the whole logical sequence of events is publication. Publication brings the work of the individual scientist under the scrutiny of the community at large, allowing it to be criticized constructively for the improvement of both the individual and the science as a whole.

The purpose of this lab exercise is to introduce you to what the scientist calls a journal. To a scientist, the term "journal" means a periodical publication where scientists report their own findings. Some journals contain only this sort of thing while others include review articles, book reviews, and news and commentary about science and scientists. In this exercise, you will get a chance to explore some of a scientist's own writings and hopefully gain some insight into his attitude toward work and the way in which he approaches it.

There are many articles to be found in newspapers and various magazines written by non-scientists about science and scientists. Many times they are very accurate but equally often they bear little resemblance to the work, which they are purported to portray. The intent and purpose of the article is often quite different from the original and further blurs the significance of both the work and the worker.

In your investigations for this lab and the subsequent library work that will be required as you go along you should be able to develop some measure of sophistication toward the various writings that deal with science, both original and reported. This should allow you to bring to your reading in future years a critical appraisal of commentary of this nature.

As with any other endeavor, it is necessary to get down into it and work. Certainly you have seen and been subjected to many, sometimes flashy, presentations of science by non-scientists on television, and in newspapers and magazines. Since you already have considerable experience along these lines, it is necessary only to examine the writings of scientists for you to be able to initiate comparisons on your own.

In the library you will find both kinds of periodicals. In some instances it is necessary to read several articles dealing with the real thing or one where you are getting some journalist's whimsical interpretation of it. The first part of the exercise is to determine which is which in our library.

When you are able to make an informed comparison about the quality of the articles presented, you should examine in detail a number of those that are the product of original scientific thought.

To do this, first write down the full reference for the article. The reference must include the following:

1. Author or authors
2. Title of article
3. Name of Journal, Volume, Number, Pages, and date of publication

Here is an example of a complete reference:
Shaman, Jeffrey, Day, Jonathan F., and Stieglitz, Marc. Drought-Induced Amplification of *Saint Louis Encephalitis Virus,* Florida. *Emerging Infectious Diseases,* Vol 8, No. 6, June, 2002, pp. 575–580.

Next, write an abstract, of the article. An abstract is a short condensation of the article describing its highlights and pertinent details. Your abstract should be one paragraph, about 100–200 words. Once this initial appraisal has been made, you should settle back and contemplate the article as an endeavor of a human being. After some thought, you should include a separate paragraph that is a secondary appraisal of the article in terms of the attitude of the person, his personality, and aims in producing the article. This section could also include your interaction with the article in terms of its impact and pertinence of life as viewed by your position.

This type of appraisal (basic data, abstract, and secondary appraisal) should be attempted for four articles that pertain to biology. At least one of these should be a review article such as those found in *Scientific American* and in the beginning pages of *Science*. Review articles are secondary sources of information. They are compiled from the original research articles published by many different scientists on a common subject. The other remaining articles must be original research. Original research is primary information. These articles are written by the researchers who are actually doing the experimental work.

Here are some points to remember for THIS assignment.

1. You must do a total of **FOUR** articles: Your instructor will inform you about the required number of review (secondary) vs. original research (primary) articles. You should choose research articles describing experimental research. You should NOT pick an article that is merely a statistical analysis of medical records, case histories, etc., or an analysis of results from a survey or questionnaire.

2. All articles must be from approved, peer-reviewed scientific journals in a field of Biology. There is a list of approved journals on the library website under Course Assignments. Your instructor will inform you if there are variations or limitations to the list of approved journals. Most biology journals on Science Direct are acceptable. Your instructor will determine how many articles may be taken from electronic databases or the Internet, and how many must be from print sources.

3. You may not use any journal more than once.

4. The article must be on a biology topic.

5. **Your instructor may require you to have your articles approved before you submit your abstracts.**

6. You must include a FULL citation of the article on the abstract that you submit.

7. You must submit a full copy of the article with your abstract stapled to it.

8. Your abstract should be short. It is not a report or a full summary. It should be about 100–200 words.

9. All abstracts must be typed, in a standard 12 pt font, such as Courier, Arial, or Times New Roman—no fancy, italic, gothic, or handwritten fonts, etc. If you single-space, it should fit on one page. If you double-space it may run into a second page.

10. Your instructor will assign a due date for the articles. The individual instructor will set the criteria for accepting late submissions. Some instructors will not accept late submissions. If a late submission is accepted, there may be penalties involved.

11. **The reference librarians will help you, but if you have any questions about this assignment, you should speak with your instructor.**

Name _____ Section _____ Date _____

Exercise 2, Part 3 — Information Literacy—Writing Assignment

PART III

The purpose of science is to develop new knowledge. This can be achieved by describing natural or man-made phenomena; by predicting if X happens, then Y will occur; by controlling phenomena so that if one controls or manipulates one variable, then we can predict a particular outcome; or finally, we can develop new knowledge by explaining phenomena. Explaining phenomena encompasses the other three methods and can lead to the development of scientific theory.

It is the goal of this assignment to have you behave as a scientist. Toward that end, you are asked to propose an experiment using two groups, conduct a brief review of the literature about your topic of interest, propose the methods for conducting the experiment, and describe the information you will collect to draw conclusions about your experiment.

In this assignment, you will propose a small research project without actually conducting the experiment. You will be asked to review the scientific literature using at least four research articles on your selected topic. As you prepare your paper, you will be expected to include the following headings:

I. Statement of the problem, question, or hypothesis

In this section, you state the research problem, question, or hypothesis. Many times research begins because of some observation and/or measurement. In your own experiences you may have made observations and wish to conduct an experiment to determine if your observations and proposed explanations are supported by evidence.

For example: In Campbell, *Biology* (6th ed.), the following hypothesis is provided: "Acid precipitation inhibits the growth of *Elodea*, a common freshwater plant." A hypothesis is testable and the variables are measurable.

In this section, you should put the problem in a historical context. What is the background to your proposed problem, question, or hypothesis? What observations have you made? Why might your proposed study be important?

II. Context of the problem, brief review of the literature

In this section, you will summarize the four scientific articles you have read to put your problem or hypothesis in perspective. This summary, or précis, provides the larger picture of the subject or field on which your research question or hypothesis is based.

The review of the literature provides some evidence that exists to support the importance of your study. The literature also provides knowledge of previous work that relates to your project and should be in agreement with your hypothesis.

For example: Campbell proposes that you design a controlled experiment to test the hypothesis that acid precipitation inhibits the growth of *Elodea*, a common freshwater plant.

A review of the literature should support the hypothesis as you learn about the effects of acids and pH on the chemical processes of cells.

Finally, the article summaries (précis) should:

a. State the main idea of the article. Then briefly describe what the author did to defend his/her work.

b. Express the original completely and cogently without personal interpretation. Do not copy one sentence from the article. Be clear and concise, using your own words.

    c. State the name of the author or article. Do not write, "In this article . . . " instead, write "Campbell notes that . . . ."

III. List of materials

In this section you simply list the materials and supplies you would need to conduct your study.

IV. Methods and procedures

How do you propose to conduct your experiment? Describe the selection of your control and experimental groups. What sample size would you use? Is the sample size practical in order to conduct your experiment? Be sure to describe in what way the experimental group would be different from the control group. Are your proposed methods feasible?

V. Findings

What information (data) would you expect to gather from your experiment? Would this information be useful to determine if your hypothesis can be supported or rejected?

VI. Conclusions

Based upon your brief review of the literature, do you have any expectations for the outcome of your experiment? You may express your opinion as to whether or not you believe your hypothesis will be supported by the evidence.

VII. Bibliography

Be sure to use the proper format for listing the references (articles) you used to prepare this paper.

This paper is worth _____% of your final grade. It must be typed and double-spaced. Your topic and proposed hypothesis must be approved before you proceed to step II.

Name _____ Section _____ Date _____

Exercise 3   Scientific Vocabulary

EXERCISE 1

Using the list provided, define the following terms:

a. epidermis
b. genocide
c. telescope
d. anthropology
e. psychology
f. microscope
g. pericardium
h. bacteriocide
i. symbiosis
j. cardiology

EXERCISE 2

Match the following terms:

_____ 1. The tube that carries the egg
_____ 2. To carry across from one place to another
_____ 3. A chemical that is water producing
_____ 4. Excessively active
_____ 5. Another term for sugar
_____ 6. To lead or carry toward a given position
_____ 7. The gel within a cell
_____ 8. Insects are joint-footed
_____ 9. Having a false body cavity
_____ 10. Dissolving by using water
_____ 11. A cell eater
_____ 12. A green formed body
_____ 13. Sound sensitive
_____ 14. Interior portion of a seed
_____ 15. First animal

a. photoreceptor
b. endosperm
c. pseudocoelum
d. chloroplast
e. cytoplasm
f. transfer
g. phagocyte
h. protozoa
i. hydrogen
k. glucose
l. oviduct
m. hypotonic
n. hydrolysis
o. hyperactive
p. afferent
q. arthropods

EXERCISE 3

In the following paragraph, underline all of the words that are of Greek or Latin derivation and answer the following questions.

Dr. Drew is a cytologist. He has been studying the effects of phototropism on the production of glucose by chloroplasts. He has been able to study his plants by using a high powered microscope. From this he was able to use a photometer and directly see the effects.

15

He put his results on a chromograph and discovered that negative phototropism produces sublevels of glucose and, therefore, slows down the process of autotrophism.

a. What is Dr. Drew's profession?
b. Why does he use a microscope to study the plants?
c. How does he measure the amount of light?
d. How does he map out the effect of light on color changes?
e. Would positive phototropism improve the production of glucose?

EXERCISE 4

Fill in the blank with the correct word.

1. This word describes something that can be seen with the naked eye. _____
2. This word describes the quality of having many colors. _____
3. This word describes the self-digestion or breaking apart that occurs in some cells. _____
4. This word means "to cut out." _____
5. This word means "the science of measuring the human body (man) and its parts." _____
6. This word describes any or a variety of substances which inhibit the growth of or destroy microorganisms. _____

EXERCISE 5

Break apart and underline the parts of the word and give the appropriate meaning.

Example: sym/bio/sis   *sym = with, bio = life*   *e/ffer/ent*   *e = away, fer = carry*

1. photometer
2. ectoplasm
3. phagocytosis
4. morphology
5. circumspection
6. fragmentation
7. telephone
8. endodermis
9. circumflex
10. translucent
11. autotrophic
12. gastrectomy
13. photosynthesis
14. oviduct
15. spermatogenesis

LATIN ROOTS

1. aqu = water
2. capit = head
3. corp = body
4. duct, duc = lead, tube
5. fer = carry, bear
6. ject = throw
7. spec = look
8. solv, solut = free, loosen
9. cid, cis = cut, kill
10. flex, flect = bend, twist
11. frag, fract = break
12. luc = light
13. omni = all
14. sequ, secut = follow
15. son = sound
16. tact = touch
17. tend, tens, tent = touch
18. tenu = thin
19. tract = draw, pull

PREFIXES

1. ab, abs = away
2. ad = to, toward
3. af = to or toward
4. ante = before
5. brachio = arm
6. circum = around
7. con = with, together
8. de = down, away from
9. dent, denti = tooth
10. ex, e, ef = out
11. inter = between, among
12. intra = within
13. ovi, ovo = egg
14. per = through
15. post = after, behind
16. pre = before, in front of
17. semi = half
18. sub, sus = under, below
19. super = over
20. trans = across

GREEK PREFIXES

1. a, an = without
2. andro = man, male
3. anthro = joint
4. anti, ant = against
5. auto = same, self
6. bio = life
7. car, cardio = heart
8. chloro = green
9. chondro = cartilage
10. coela, coelo = cavity, hollow
11. cyto = cell
12. dont = tooth
13. ecto = outside
14. endo = within
15. entero = intestine
16. epi = to, on, over, against
17. gastro = stomach
18. gluco, glyco = sweet
19. hyper = over, excessive
20. hypo = under, below, less than
21. iso = same
22. macro = large
23. micro = small
24. mono = single
25. morph, morpho = shape form
26. peri = surrounding
27. proto = first
28. pseudo = false
29. poly = many
30. sperma, spermato, spermo = seed
31. sym, syn = together, within

GREEK ROOTS AND OTHER PREFIXES AND SUFFIXES

1. anthropo = man
2. ectomy = to cut out
3. phagus = eater
4. meter = that which measures, length
5. logy = study of
6. lysis, lytic, lye = loosening, breaking apart
7. graph = map
8. gen = birth, race, kind
9. photo = light
10. psych = mind
11. hydra = water
12. phyte = plant
13. phon = sound
14. plasm, plasmo, plast = form, mold, gel
15. chrom = color
16. scop = see
17. tele = far, distant
18. derm, dermis = skin
19. zoa = animal
20. trop = turning
21. troph, tropho = feeding, nourishing

Part II

Chemistry

Name _____ Section _____ Date _____

Exercise 4 Laboratory Measurements

OBJECTIVES

After completing this exercise, you should be able to:

1. List, define, and identify by symbol the following:
    a. three basic units of the metric system
    b. four metric prefixes that make the basic units smaller
    c. three metric prefixes that make the basic units larger
2. Measure a liquid with a graduated cylinder and/or pipette, noting largest and smallest amounts of liquid to be measured by each.
3. Measure the mass of an object using a balance, noting largest and smallest units on the balance.
4. Discuss the relationship between a milliliter of water and a gram of water.

PRELIMINARY INFORMATION

The United States is the only major country in the world that has not adopted the International System of Units or metric system. This creates problems in international trade and other areas. It is apparent the world is not going to convert to our system, anyway. Why should they? The system of measurements we use is clumsy. A conversion from one unit to the next is tedious and time-consuming, whereas conversion within the metric system is done by simply moving a decimal point. It works the same way our money system works.

It is apparent that metric conversion has already begun. Many foods, like soda, are packaged with metric equivalents on them, skis are sold in metric sizes, and even some highway signs give distances in kilometers.

The scientific community adopted the metric system in 1960. All measurements used in this lab book will be in the metric system.

PART I BASIC UNITS

Useful Conversions to Remember

1 millimeter = thickness of a dime
1 centimeter = width of an aspirin
1 meter = width of a doorway
1 gram = mass of a dime
100 grams = 1/4 pound of candy
1 kilogram = slightly more than two pounds of hamburger
1 liter = slightly more than one quart of beer

Basic Metric Units

Unit	Symbol	Measures
meter	m	length
gram	g	mass (weight)
liter	L	volume

From *Biological Investigations: Lab Exercises for General Biology* 11th edition by Edward Devine. Copyright © 2004 by Kendall/Hunt Publishing Company. Reprinted by permission.

Metric Prefixes

Prefix	Symbol		Meaning
Makes Smaller			
deci-	d	decimal system with divisions of 10	one-tenth of or 0.1
centi-	c	cent is one-hundredth of a dollar	one-hundredth of or 0.01
milli-	m	a millipede has 1000 legs	one-thousandth of or 0.001
micro-	μ	a microscope sees small things	one-millionth of or 0.000001
Makes Larger			
deca-	D	Greek word for 10	ten times or 10
hecto-	h	Greek word for 100	hundred times or 100
kilo-	k	kilocalorie is 1000 calories	thousand times or 1000

PART II ABBREVIATIONS

Complete the following:
Using combinations from the preceding table, define the following symbols:

dm _____   kg _____

mm _____   mg _____

mL _____   Dg _____

cm _____   km _____

PART III UNITS OF LENGTH

The basic unit of length in the metric system is the meter. Larger or smaller units of measure are obtained by either multiplying or dividing the meter by 10 or a factor of 10 such as 100 or 1000.

The following diagram expressed the relationships between the standard units in the metric system.

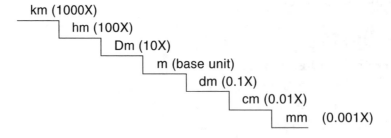

Figure 4.1
Metric Hierarchy

Conversion: Two methods can be used in converting one unit of measurement to another.

1. Move decimal point.

   To use this method, you must know the stair step relationship in the metric hierarchy. To convert **when moving down the steps, move the decimal point to the right for each step. When moving up the steps, move the decimal point to the left.** To help remember the steps, use a saying such as "Kids Have Dropped Back Dead Converting Metrics." "Back" represents the base unit.

2. Factor method

   Numerically, conversions of many kinds can be done by using the factor method. This involves a simple concept: Any number multiplied by 1 is still the same number. To measure a distance this concept would be: any distance multiplied by 1 is still the same distance (however the units may change). For example, 1 m equals 100 cm equals 1000 mm. When actually using this concept, the number 1 is not actually used but rather a factor or fraction equal to one. For example, using the relationships listed above, several factors or fractions can be made.

$$\frac{1\,m}{100\,cm} = \frac{100\,cm}{1\,m} = \frac{1\,m}{1000\,mm} = \frac{1000\,mm}{1\,m}$$

A conversion factor is a fraction that has the following characteristics:

1. The numerator is equivalent to the denominator (which makes the factor equal to 1).
2. Put the unit you want for an answer in the numerator.
3. Put the unit you want to eliminate in the denominator.

To convert any measurement, use this method:

$$\text{Measurement} \times \text{Conversion factor} = \text{Answer}$$

Example 1

Convert 75 cm to m.

$$75\,cm \times \frac{1\,m}{100\,cm} = .75\,m \text{ or move the decimal point two places to the left.}$$

Note: Arrange the conversion factor so the units in the starting measurement will cancel the same units in the denominator of the conversion factor. Otherwise your answer will also have nonsense units.

Example 2

Convert 15.4 cm to mm.

$$15.4\,cm \times \frac{10\,m}{1\,cm} = 154\,mm \text{ or move the decimal point one place to the right.}$$

Remember: as the units get larger, as when converting centimeters to meters, the numbers get smaller.

Directions

1. Obtain a meterstick and measure the following:
   a. Measure the length of the lab table. It is _____ mm, _____ cm, or _____ m.
   b. Measure the width of the lab table. It is _____ mm, _____ cm, or _____ m.
   c. Measure the width of your hand. It is _____ mm, _____ cm, or _____ m.
   d. Measure the length of your little finger. It is _____ mm, _____ cm, or _____ m.
   e. Measure the width of your little finger. It is _____ mm, _____ cm, or _____ m.

PART IV UNITS OF VOLUME

Liquids are most accurately measured by using graduated cylinders or pipettes. Pipettes are used for small amounts of 10 mL or less, whereas graduated cylinders are used to measure 10 mL to 1000 mL.

1. Locate the demonstration of various graduated cylinders containing colored water. Position your head so the graduated cylinder is at eye level. Notice that the water forms a depression called a *meniscus*. Measure from the bottom of this depression. Note the following:

Cylinder	A (Largest)	B	C	D	E (Smallest)
Value of graduations: (amount between two of the smallest lines)	___	___	___	___	___
Total capacity that can be measured:	___	___	___	___	___
Amount of colored water present:	___	___	___	___	___

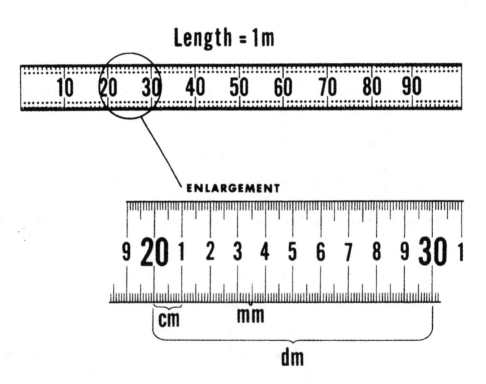

Figure 4.2
Meterstick

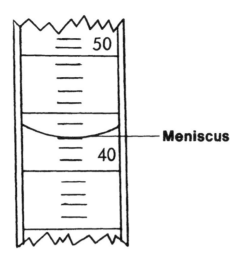

Figure 4.3
Graduated Cylinder Shows 43 mL

2. Locate the demonstration of various pipettes.

   Pipettes are long glass straws and will be marked or calibrated differently. At the end of many pipettes will be notations such as *5 in 1/10* or *10 in 1/100*, which means the pipette will measure a total of *5 mL in units of tenths* of a mL or a total of *10 mL in units of hundredths* of a mL. If such a notation is not present, you will have to look at the pipette itself and figure it out.

   a. Select any three different pipettes and note the following:

Pipette

	A (small)	B (medium)	C (large)
Total capacity that can be measured	_____	_____	_____
Value of graduations: (amount between two of the smallest lines)	_____	_____	_____

   b. To practice pipetting, obtain a Pasteur pipette, a beaker, and a 10 mL graduated cylinder. Determine how many drops of water are in 1 mL. Record here _____.

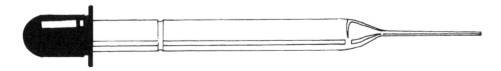

Figure 4.4
Pasteur Pipette

PART V UNITS OF MASS

1. Locate the demonstration of balances and identify all of the labeled parts.

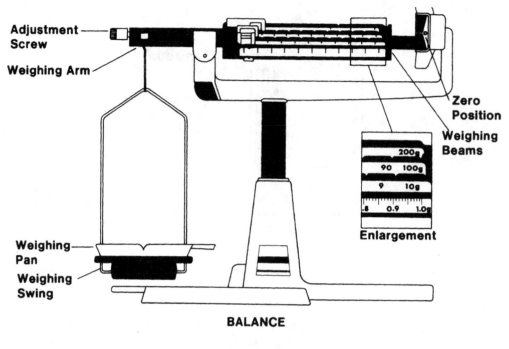

Figure 4.5
Balance

2. All weights on the calibrated beams should be at zero. The weighing beams should read zero. If they do not, first check if the number on the bottom of the weighing pan is the same as the number on the bottom of the weighing swing. If the numbers do not match, locate the correct weighing pan. If the numbers do match, then the balance can be zeroed by using the adjustment screw.

3. Most scales will have four beams. Starting from the front beam and proceeding to the back, note the following:

	1st Beam	2nd Beam	3rd Beam	4th Beam
Total mass measured:	___	___	___	___
Smallest mass measured: (Zero is not mass.)	___	___	___	___

4. What is the maximum capacity of the total balance?. _____
5. What is the smallest amount that can be measured? _____
6. Place a small beaker on the balance and weigh it. Start with the farthest beam and proceed forward until the final weight has been determined. Record here _____.

PART VI DENSITY

Density is the relationship between the mass and the volume of a substance. For liquids and solids, the density is expressed as the number of grams in 1 mL. For gases, the density is the number of grams in 1 liter. The formula for this relationship is:

$$\text{Density} = \frac{\text{Mass}}{\text{Volume}} \quad or \quad D = \frac{M}{V}$$

In this exercise, the density of water will be calculated.

You will determine the weight of 10 mL of water, then divide the weight by 10 to determine the density (which is the mass of 1 mL).

1. Fill a small graduated cylinder (not a beaker) to the 10 mL mark.
2. Weigh a small beaker on the balance and record the weight in the proper space below.
3. Pour the 10 mL of water into the beaker, weigh again, then record in the proper space.
4. Calculate the weight of the water.

   Weight of beaker and water   _____

   Weight of beaker   _____

   Weight of water   _____

5. Calculate the density of water in the space below by dividing the weight of the water by the volume of 10 mL.

   Density of water = _____

6. Ideally the density of water should be 1 g/mL.
7. If your data showed a different density for water, list some of the possible sources of error here.

PART VII ADDITIONAL MEASUREMENTS

1. Using a balance, determine the weight of the following items from the glassware drawer.

   a. Test tube = _____ g
   b. Microscope slide = _____ g
   c. 250 mL beaker = _____ g
   d. 50 mL beaker = _____ g. *Note:* Keep the beaker for use in the next exercise.

2. Using the 50 mL beaker from the previous exercise and a graduated cylinder, put 50 mL of water in the 50 mL beaker. Determine the mass of the water and the beaker and record here. _____ Use this data to calculate the density of water. Carefully show all calculations.

3. What is the length of this line?

   |_____| _____ mm = _____ cm

Exercise 4 Laboratory Measurements   27

4. What is the length of this page _____ mm = _____ cm
5. Complete the following conversions:
   a. 0.33 m = _____ cm = _____ mm
   b. 0.78 liters = _____ mL
   c. 350 mg = _____ g
   d. 454 mm = _____ cm = _____ m
   e. 5.2 kg = _____ g = _____ mg
   f. 1350 mL = _____ L
   g. 550 g = _____ kg
   h. 2.8 L = _____ mL
   i. 9342 mm = _____ cm = _____ m
   j. 0.03 L = _____ mL
   k. 2.465 m = _____ cm = _____ mm
   l. 0.674 kg = _____ g = _____ mg
6. Use the density table to help solve the following problems.

Density Table

Substance	Density (g/mL)
Ethyl alcohol	0.79
Water	1.00
Sugar	1.6
Carbon Tetrachloride (a liquid)	1.6
Aluminum	2.7
Silver	10.5
Lead	11.3
Mercury	13.5
Gold	19.3

   a. If 15 mL of a liquid has a mass of 11.85 grams,
      (1) what is the density? _____
      (2) what is the liquid? _____
   b. A beaker with a mass of 30 grams has 40 mL of liquid poured into it. The beaker and the liquid have a mass of 94 grams.
      (1) What is the density of the liquid? _____
      (2) What is the liquid? _____
   c. A prospector drops a 192-gram piece of solid material into a graduated cylinder containing 50 mL of water. The water level rises to 67 mL. What is the solid? _____ Show calculations.

28   Exploring Biology in the Lab

Name _____ Section _____ Date _____

Exercise 5 | Measurement

OBJECTIVES

After completing this exercise, you should:

1. Be familiar with and able to use measuring devices.
2. Be able to make simple conversions between the British and metric systems.
3. Be acquainted with scientific notation and significant figures.

EQUIPMENT

1. Charts for definitions, abbreviations, conversion factors, powers of 10 in decimal form, and scientific notation.
2. Rulers:
   Metric—large and small
   British
3. Graduated cylinders 100 ml, 10 ml, 5 ml
4. Pipettes—5 ml

EXERCISE 1

Instructions: abbreviate the following:

a. centiliter
b. milligram
c. kilogram
d. kilometer
e. centigram
f. decigram

EXERCISE 2

Instructions: given scientific notation:

1. Write out the following terms in complete decimal form:
    a. 5.24×10^{-2}
    b. 5.24×10^{0}
    c. 5.24×10^{-3}
    d. 5.24×10^{3}
    e. 5.24×10^{1}
    f. 5.24×10^{-1}
2. Write the following decimals in scientific notation:
    a. 0.00367
    b. 0.0367

c. 0.000367
   d. 54.0
   e. 27.1
   f. 7,200,000
3. Change the following to scientific notation:
   a. 4312
   b. 2100
   c. 53,958
   d. 60
   e. 6010
4. Write out the following numbers:
   a. 4.1×10^0
   b. 3.57×10^2
   c. 1.1×10^2
   d. 4×10^{-1}
   e. 8.9×10^{-1}

EXERCISE 3

Instructions: Using scientific notation, calculate the following:

a. $390 \div 130$
b. $100,000 \div 100$
c. 4100×100
d. 2700×20
e. $657 + 300 + 950$
f. $28 + 4500 + 1100$
g. $7900 - 950$
h. $94,000 - 2,600$

EXERCISE 4

Instructions:

1. Place 5 ml of water in a 100 ml graduate cylinder. Pour this into a 10 ml cylinder and finally into a 5 ml cylinder.
   a. Does the amount measure out correctly?
2. Using the 5 ml cylinder, measure out 5 ml then pour it into the 10 ml cylinder and then into the 109 ml cylinder.
   a. Does the amount measure out the same?
   b. Do you think it's better to measure out a 5 ml amount in a 5 ml cylinder or a 100 ml cylinder for accuracy? Why?
3. Using the 5 ml pipette, pipette 5 ml of water into a 5 ml cylinder.
   a. Do they measure out the same?
   b. Of all, which measures out most accurately?

EXERCISE 5

Instructions: Calculate the following:

a. How many cm in 400 m?
b. How many mm in 85 m?
c. How many m in 600 mm?
d. How many m in 7500 cm?
  (a) Measure your height: How many meters? _____ cm? _____ mm? _____
  (b) What is the length of the desk top in m? _____ cm? _____ mm? _____

EXERCISE 6

Instructions

1. Go out and collect 20 of the same type of leaves and measure their length. Fill in the chart. Using graph paper, prepare a graph showing the length of each leaf. Label the X-axis for leaf length and label the Y-axis for the leaf number.

Leaf #	Length
1.	_____
2.	_____
3.	_____
4.	_____
5.	_____
6.	_____
7.	_____
8.	_____
9.	_____
10.	_____
11.	_____
12.	_____
13.	_____
14.	_____
15.	_____
16.	_____
17.	_____
18.	_____
19.	_____
20.	_____

2. Collect 20 different types of leaves and then measure the length of 20 different leaves. Fill in the chart. Using graph paper, prepare a graph showing the length of each leaf. Label the X-axis for leaf length and label the Y-axis for the leaf number.

Leaf #	Length
1.	_____
2.	_____
3.	_____
4.	_____
5.	_____
6.	_____
7.	_____
8.	_____
9.	_____
10.	_____
11.	_____
12.	_____
13.	_____
14.	_____
15.	_____
16.	_____
17.	_____
18.	_____
19.	_____
20.	_____

QUESTIONS

1. Calculate the average of both samples.

2. What does your data tell you about leaf variation in leaves of the same species versus leaves of different species?

3. Are there other variables to be considered?

Name _____ Section _____ Date _____

Exercise 6 Enzyme Activity *(Protein)*

3D form

INTRODUCTION

Enzymes are biological **catalysts** Like all catalysts, they speed up chemical reactions. They accomplish this by **lowering the activation energy** needed for the reaction to proceed. Almost all biochemical reactions and pathways in your cells are controlled by enzymes. Nearly all enzymes are proteins and proteins have very complex biological shapes. As an enzyme is formed, the amino acid chain folds up to produce a pocket or groove on the surface of the enzyme with a unique shape. This pocket is called the **active site.** When an enzyme catalyses a chemical reaction, it binds one of the reactants in the active site. The substance that binds at the active site is called the **substrate.** The enzyme and substrate form a complex **(enzyme-substrate complex)** that makes it easier to carry out the chemical reaction. At the end of the reaction, the products are formed, and the enzyme is released, unchanged, to be used over again. Enzymes and substrates are specific for one another because the shape of the substrate matches the shape of the active site. This is sometimes called the "lock and key" theory. As proteins, enzymes are very sensitive to conditions of temperature and pH. The laboratory today will examine the effects of temperature, pH, and inhibitors on the plant enzyme, **catecholase,** extracted from a potato. The substrate, **catechol,** is colorless. When mixed with the enzyme, catechol is converted to a brown compound called benzoquinone. You can detect this reaction by the formation of a brown color.

PROCEDURE

1. Collect a few ml of well-mixed enzyme extract from stock bottle into a small beaker. Label beaker.
2. Collect about 200 ml of substrate solution (from stock bottle) into a larger beaker. Label beaker.
3. Fill *third* small beaker with distilled H_2O. Label beaker.
4. Place *separate* pipets with bulbs in each solution.
5. **Controls**
   a. Place 5 ml of water in a test tube. Add 10 drops of substrate. Mix vigorously to introduce oxygen. Observe for 1–2 minutes. Does any color develop? Why is this tube called the negative control? *Remain clear.*
   b. Place 5 ml of water in a second test tube. Add 10 drops of substrate and 10 drops of enzyme. Mix vigorously to introduce oxygen. Observe for 1–2 minutes. Record the time it takes for color to develop. What color develops? Why is this tube called the positive control? *2 min yellowish.*

[Catechol] → catecholase → [Benzo Quinone]
Reactant / substrate — enzyme

Qualitative Result = color Δ

6. **Experimental Procedure A: Effect of temperature on the enzyme**
    a. Fill 3 clean test tubes about one-third full with water. It is important that all 3 tubes be filled equally. Add 10 drops of substrate to each tube. Label tubes #1S, #2S, and #3S.
    b. Prepare 3 more clean test tubes and fill about one-third full with water. Add 10 drops of enzyme. Label tubes #1E, #2E, #3E.
    c. Expose 1 test tube of substrate and 1 test tube of enzyme to each of the following conditions.

    SET 1—Ice Bath (10°C)   SET 2—Body temperature (37°C)
    SET 3—Boiling Water Bath (90°C)

    Hold for 15 minutes in each condition.

    Note: Cap the tubes held at body temperature to prevent spilling.

    d. Following exposure, **quickly** mix the enzyme test tube with the substrate test tube and agitate the contents. **(Note: Add enzyme to ~~starch~~ substrate solution).**
    e. Does any product form in any of the test tubes? Record the time it takes for product to form in each of the test tubes. Can you draw any conclusions about the effect of temperature on the enzyme?

7. **Experimental Procedure B: Effect of pH on the enzyme**

    Note: Be careful when handling acids and bases. If you spill any solution on your skin, wash your hands immediately.

    a. Obtain 3 clean test tubes. Fill one about half full with an acid solution. Fill the other about half full with a basic solution. Label the tubes "A" for acid and "B" for base. Fill the last tube about half full with distilled water. Label this tube "N" for neutral.
    b. Add 10 drops of enzyme to each tube. Let the enzyme sit in the tubes for 10 minutes.
    c. After the 10 minutes, immediately add 10 drops of substrate to each tube.
    d. Observe for product formation and record the time for product to form. Can you draw any conclusions about the effect of pH on the enzyme activity?

8. **Experimental Procedure C: Effect of an inhibitor (Phenylthiourea) on the enzyme**

    Note: Be careful when handling Phenylthiourea, because it is a poison. You should wear gloves.

    a. Obtain two test tubes. Fill one half full with distilled water. Fill the other half full with phenylthiourea.
    b. Add 10 drops of enzyme to each tube. Label the tubes. Let the enzyme sit in the tubes for 10 minutes.
    c. After the 10 minutes, immediately add 10 drops of substrate to each tube.
    d. Observe for product formation and record the time for product to form. Can you draw any conclusions about the effect of PTU on the enzyme activity?

9. **Experimental Procedure D: Effect of substrate concentration on the reaction**
    a. Obtain 3 test tubes. Fill each half full with distilled water.
    b. Add 5 drops of substrate to one test tube. Add 15 drops of substrate to the second tube. Add 30 drops of substrate to the third tube. Label the tubes.
    c. Immediately add 10 drops of enzyme to each tube.
    d. Observe for product formation and record the time for product to form. Can you draw any conclusions about the effect of substrate concentration on the rate of the reaction?

10. **Experimental Procedure: Effect of enzyme concentration on the reaction**
    a. Obtain 3 test tubes. Fill each half full with distilled water.

b. Add 10 drops of substrate to each tube.
c. Add 5 drops of enzyme to one test tube. Observe for product formation and record the time for product to form.
d. Add 10 drops of enzyme to the second tube. Observe for product formation and record the time for product to form.
e. Add 20 drops of enzyme to the third tube. Observe for product formation and record the time for product to form.
f. Can you draw any conclusions about the effect of enzyme concentration on the rate of the reaction?

Experiment A: Effect of Temperature

	Time (sec) to Product Formation	Comments
Ice Bath (10°C)	—	no Δ
Body Temperature (37°C)	2 sec	yellowish
Boiling Water Bath (90°C)	—	no Δ
Experiment B: Effect of pH		
Acidic	—	no Δ
Neutral	4 sec	yellowish
Basic	6 sec	greenish
Experiment C: Effect of Inhibitor		
Enzyme + Water		yellow
Enzyme + PTU		clear
Experiment D: Effect of Substrate Concentration		
5 drops	—	pale yellow
15 drops	*Feedback Inhibition*	richer yellow } same color
30 drops		richer yellow
Experiment E: Effect of Enzyme Concentration		
5 drops	45 sec	pale yellow
10 drops	35 sec	pale yellow } with time deeper
20 drops	25 sec	pale yellow

Name _____ Section _____ Date _____

Exercise 7 — Food Nutrient Analysis (Organic Molecules)

OBJECTIVES

After completing this exercise, you should be able to:

1. Understand and demonstrate biochemical tests for the presence of various organic molecules.
2. Use these tests in a controlled experiment to identify components of an unknown compound.
3. Write a lab report explaining the procedures and results.

In each of the following tests, you will use a pair of controls and one unknown solution. The controls are known solutions, one containing the substance of interest (the positive control) and one without it (the negative control). These controls will help you interpret the results of your tests on the unknown.

IODINE TEST FOR STARCH (shake each time)

Background: Starch is a polymer of glucose with a coiled shape that reacts with iodine to form a blue-black color. Iodine does not react with differently shaped carbohydrates; it remains yellowish-brown.

Procedure: Place 5 ml of water and 4 drops of iodine solution to each of three labeled test tubes. Add the following:

Tube 1—10 drops of water N
Tube 2—10 drops of starch solution P
Tube 3—10 drops of unknown U

Record any changes in color.

INDOPHENOL TEST FOR VITAMIN C

Background: Vitamin C, ascorbic acid, is a strong reducing agent that bleaches a blue solution of indophenol, turning it colorless.

Procedure: Place 20 drops of indophenol solution in each of three labeled vials. Add the following:

Tube 1—3 drops of water
Tube 2—3 drops of Vitamin C
Tube 3—3 drops of unknown

Shake the vials and record any color changes. If no change takes place, add more of the respective solutions, 1 drop at a time until there is a change.

BENEDICT'S TEST FOR REDUCING SUGARS

Background: Many monosaccharides contain free carbonyl (C=O, an aldehyde or a ketone) groups that can reduce certain metallic ions. Benedict's reagent, a blue-colored alkaline solution of cupric sulfate ($CuSO_4$), contains such an ion. In the presence of high pH and heat, the cupric ions (Cu^{++}) of the reagent are reduced by the sugar to cuprous ions (Cu^+) which then form the red-orange compound cuprous oxide (Cu_2O). (Small amounts of reducing sugars produce an intermediate greenish color.)

Procedure: Place 5 ml of water and 20 drops of Benedict's solution into each of three test tubes, labeled 1–3. Add the following:

Tube 1—10 drops of water

Tube 2—10 drops of sugar solution

Tube 3—10 drops of unknown

Record the color of each solution. Carefully place the tubes in the hot water bath for 5 minutes, then remove them and record the color changes.

BIURET TEST FOR PROTEIN

Background: Biuret reagent contains $CuSO_4$. The Cu^{++} ions in this solution, when mixed with NaOH and protein, react with peptide (C-N) bonds to form a pink or purple precipitate that settles, leaving the liquid pink to purple. If protein is not present, a blue precipitate forms and the liquid remains clear.

Procedure: Place 5 ml of water in each of three labeled tubes. Add the following:

Tube 1—10 drops of water

Tube 2—10 drops protein solution

Tube 3—10 drops of unknown

Now add 20 drops of sodium hydroxide and 20 drops of copper sulfate solution to each tube and shake gently. Record results.

SUDAN IV TEST FOR LIPID

Background: As most know, oil (lipid) and water do not mix. Sudan IV is a fat-soluble dye (which is also insoluble in water); in the presence of both fat and water, the stain will be concentrated in the fat layer, giving it a red color.

Procedure: Place 5 ml of water in each of three labeled tubes. Add the following:

Tube 1—10 drops of water

Tube 2—10 drops of vegetable oil

Tube 3—10 drops of unknown

Using a toothpick, add a small amount of Sudan IV to each tube. Cover the end of each tube with your thumb and shake for 5 seconds, then leave the tubes undisturbed for 3 minutes. Record the results.

QUESTIONS

Include answers to the following in the Discussion section of your lab report.

1. What are the controls in each test and why are they included?
2. How could you determine the form of carbohydrate stored in a plant?
3. What results would you expect in the Biuret test from a solution of the free amino acid glycine? Why?
4. What results would you predict when testing table sugar (sucrose) with Benedict's solution? Why?
5. What is a vitamin? Would you expect the indophenol test to detect other vitamins? Why?
6. What limitations or possible problems can you see in the tests you learned today?

Food Nutrient Lab
Name of Test

Unknown G (Karo syrup)

Material Tested

	Sugar	Starch	Lipid	Protein	Vitamin C
Negative Control (Indicator Reagent + H₂O)	Aquamarine p̄ bath no Δ	no Δ	∅ lipid clear ring of Sudan IV	cloudy light blue	no Δ – dk blue p̄ 75 gtts started to dilute – sl. lighter in color
Positive Control (Indicator Reagent + H₂O + pure known nutrient)	Aquamarine p̄ bath greenish brown (drab olive)	black	⊕ lipids reddish top	bluish lavender c̄ sediment.	clear
Unknown " "	Aquamarine p̄ bath cloudy yellowish orange	reddish brown @ bottom fading to amber	∅ lipid clear c̄ ring of sudan IV	bluish tint clear	no Δ – dk blue p̄ 65 "G" gtts turning purplish (diluted)
Unknown " "	⊕ present	⊕ present	⊖ present		⊖ present
Unknown " "					

Note: Unknown Testing consists of specific Indicator Reagent + H₂O + specific unknown.

Know reagents + what they react with

⊕ is of more assistance.

Part III

Cells

Name _____ Section _____ Date _____

Exercise 8 | Using the Microscope

PROBLEM

How is the monocular microscope properly transported, cleaned, and used for magnification?

OBJECTIVES

After completing this exercise, you should be able to:

1. Properly transport and return the microscope to and from the microscope cabinet.
2. Clean the lenses.
3. Name and give the functions of the major parts of the light microscope.
4. Calculate the total magnification.
5. Locate and focus specimens on a microscope slide.

PRELIMINARY INFORMATION

The human eye is only capable of seeing objects the size of a grain of sand or larger. Thus the cell and its substructures that make up the basis of life are invisible to the unaided eye. The light microscope is the first step in making the "invisible world" visible.

The light microscope is a basic instrument used in the study of biology. There are two types of light microscopes. The dissecting microscope is used when low level three-dimensional magnification is desired. The compound light microscope is used to view thin pieces or sections of material placed on a slide. To improve contrast, stains are often applied to the specimen.

PART I HANDLING THE COMPOUND MICROSCOPE

1. Before picking up a microscope, understand that it is an expensive and precision instrument and should be treated very carefully. Carry the microscope in front of your body with two hands: hold the base with one hand, hold the arm with the other hand.
2. Obtain a microscope from the cabinet as directed by your instructor. Notice the number on the microscope matches the number on the space in the cabinet. Be certain to **return the microscope to its properly numbered** space when you are finished.
3. Set the microscope on the lab table, back slightly from the edge.
4. Remove the dust cover and place it out of the way.
5. Inspect the microscope. The electric cord should be looped around the eyepiece in two or three large loops. Tight coiling of the wire can eventually break the wires inside, so **DO NOT** wrap the cord tightly around the base of the microscope.
6. The mechanical stage should not have parts protruding.
7. The shortest objective on the revolving nosepiece should be pointed down.

   Note: It is very important that the microscope be returned to the proper cabinet space with all parts positioned properly.
8. You are responsible for this microscope. If it is put away improperly or damaged, you could be held responsible. Therefore, report any incorrect storage, damage, or malfunction immediately to your instructor.

From *Biological Investigations: Lab Exercises for General Biology* 11th edition by Edward Devine. Copyright © 2004 by Kendall/Hunt Publishing Company. Reprinted by permission.

PART II PARTS OF THE COMPOUND MICROSCOPE

Using Figure 8.1 and your light microscope, identify and become familiar with the functioning of the following parts:

1. **Ocular** or eyepiece: the lens that you look through when viewing a slide. This lens magnifies ten times (10×) and has a built-in pointer.
2. **Body Tube:** the tube holding the eyepiece.
3. Objectives: these are the lenses that point to the slide. Most microscopes will have three objectives. Examine your microscope. The magnification of the objective is stamped on the side. There are four standard types of objectives manufactured:
    a. **Scanning objective, 4×.**
    b. **Low power or low dry, 10×.**
    c. **High power or high dry, 40×.**
    d. **Oil immersion, 100×.** Not present on general biology microscopes.
4. **Revolving nosepiece:** this holds the objectives. It can be turned by hand to position any one of the three objectives over the slide.
    Note: Be sure the objective is properly snapped into place before trying to use the microscope.
5. **Arm:** the heavy portion supporting the body tube and the revolving nosepiece.
6. **Coarse adjustment** or coarse focus knob: the inner and larger knob on the lower portion of the arm.
7. **Fine adjustment** or fine focus knob: the outer and smaller knob on the lower portion of the arm.
8. **Base:** the part in contact with the lab table.
9. **Stage:** the flat part of the microscope on which the slide is placed.
10. **Mechanical stage:** the spring-loaded mechanism for holding and positioning the slide under the objective. The positioning knobs are on the right side of the stage.
11. **In-stage condenser:** this focuses light onto the slide.
12. **Iris diaphragm:** moving the lever opens and closes the iris diaphragm regulating the amount of light reaching the slide.
13. **Illuminator:** the built-in light source is turned on with a push switch near the illuminator.

PART III MAGNIFICATION AND ESTIMATING THE SIZE OF OBJECTS

The compound microscope has two separate lens systems. The image coming through the objective lens is magnified by the amount printed on the objective. The image is again magnified ten times (10×) as it passes through the ocular lens. **The total magnification is calculated by multiplying the magnification of the objective lens times the ocular lens.** The field of vision is the part of the slide seen through the ocular. As the magnification increases, the field of vision decreases. The actual size of an object can be estimated by knowing the diameter of the field of vision. Table 8.1 relates magnifications and fields of vision to the different objective lens.

With this information you can approximate the size of an object by estimating what fraction of the diameter of the field the object occupies. For example, if a spherical cell viewed under high power is about a third of the distance across the field of vision then the cell would be about 0.1 mm or 100 microns in diameter.

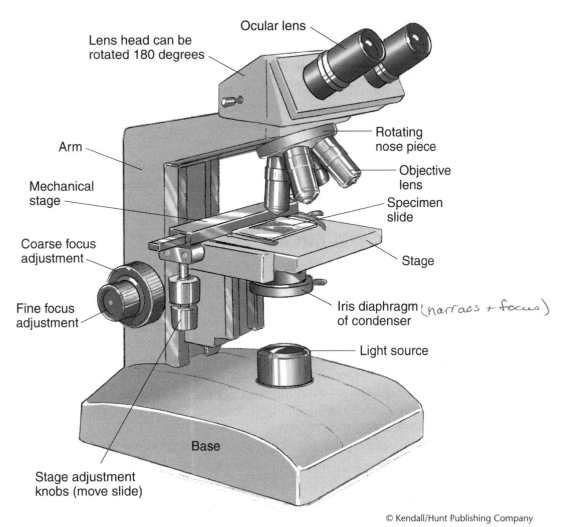

Figure 8.1
Compound Microscope

Iris diaphragm (narrows + focus)

Table 8.1
Microscope Magnifications

Objective Lens	Ocular Lens	Total Magnification	Diameter of the Field of Vision in mm	Microns
Scanner 4X Red	10X	40X	3.75	3,750
Low-power 10X Yellow	10X	100X	1.50	1,500
High-power 43X Blue	10X	430X	0.35	350
Oil immersion 100X	10X	1000X	0.15	150

Red   shortest — lowest

PART IV CLEANING THE COMPOUND MICROSCOPE

Cleaning the microscope lenses and slides before use is important for a clear, sharp image. Cleaning also prevents you from wasting time viewing dirt on the lens and slide.

1. **Only lens paper** should be used to clean the microscope lenses and prepared slides. Paper towels and facial tissue can scratch the lenses.
2. Carefully wipe the ocular and objective lenses with the lens paper.
3. Obtain a prepared microscope slide.
4. Using lens paper, clean the slide as you would clean sunglasses by applying light even pressure to both sides of the slide. Excessive pressure should never be used since the mounting medium never becomes completely hard.

PART V FOCUSING THE MICROSCOPE

1. Using the coarse adjustment knob, raise the nosepiece until it stops.
2. Put the scanning (4×) objective into the down or viewing position.
3. Examine the slide for the approximate position of the specimen by holding the slide up to the light.
4. Place the slide on the stage by moving the spring-loaded arm of the mechanical stage slightly to the side. DO NOT LIFT UP ON THE MECHANICAL STAGE ARM because it can be permanently bent. Release the spring-loaded arm against the side of the slide.

   Note: Do not put the arm over the slide, just against the side. Be sure the side of the slide closest to you is tight against the mechanical stage. This allows full view of the slide and eliminates slipping during fine movements.

5. Using the knobs on the mechanical stage, position the specimen under the objective.
6. Turn on the microscope light and open the iris diaphragm about one-fourth.
7. Using the coarse focus knob, position the objective all the way down.
8. While looking through the ocular, slowly turn the coarse focus up. When an object appears, focus in on it using the smaller fine focus knob. Adjust the light for the best image. If nothing appears, move to another spot on the slide and try again.

   Note: The image is inverted and reversed due to lens optics. Commercially prepared slides mount the specimens so they will be seen properly if there is a top/bottom or left/right orientation that is significant.

9. After the specimen is in good focus under scanning power, center it in the field of view.
10. Change to the low power (10×) objective by simply swinging that objective into the down or viewing position.

11. Adjust the fine focus and light intensity to obtain the best image.
12. Change to the high power (40×) objective by simply swinging that objective into the down or viewing position.

    Note: Good quality microscopes are parcentered and parfocal. **Parcentered** means the center of the field of vision stays essentially the same in each objective. **Parfocal** means practically no change in focus has to be made when changing objectives.
13. Adjust the fine focus and light intensity to obtain the best image.
14. Never force a gear or mechanism. If any part does not operate smoothly and easily, ask your instructor to check it.
15. The proper way to look through a microscope is with both eyes open. When you begin working on this technique, clear the lab table beside the microscope so you don't see any extraneous objects. Keeping both eyes open will be awkward at first, but you should adjust in 10–15 minutes of practice. When making drawings of specimens, you may find it useful to simultaneously look at the specimen with one eye and the drawing paper with the other eye. This technique will require a great deal of practice.

PART VI DRAWING

1. On the data sheet, identify and draw a small portion of the specimen that you have isolated on the slide.
2. When you have completed your microscope work, carefully:
   a. Remove the slide.
   b. Clean the lenses using lens paper.
   c. **Put the lowest power objective in the down position.**
   d. **Position the mechanical stage so parts are not sticking out the side.**
   e. Unplug the electrical cord and loop it in two or three large loops around the eyepiece.
   f. Cover the microscope with the dust cover.
   g. **Return the microscope to its properly numbered space** in the microscope cabinet.
   h. Clean the lab table by wiping it with wet paper towels.

Name _____ Section _____ Date _____

Exercise 9 Microscopic Techniques

Many of the observations that you will record in the biology lab will be made at the microscopic level. In order to do this you will need to use the compound microscope. This lab will introduce you to the compound microscope and some of the techniques of light microscopy.

OBJECTIVES

After completing this exercise, you should be able to:

1. Identify the parts of the compound microscope and their function.
2. Describe and demonstrate the correct way to:
   - Carry the microscope
   - Clean the lenses
   - Adjust the light
   - Focus with each objective
   - Examine prepared slides
   - Prepare and examine a wet-mount slide
   - Determine the total magnification
   - Estimate the size of objects
   - Prepare the microscope for storage and put it away

CARE OF THE MICROSCOPE

1. Obtain the microscope that is assigned to you from the cabinet. DO NOT USE ANOTHER MICROSCOPE. If your microscope is missing or there is a problem, notify the instructor.
2. Always carry the microscope in an upright position, directly in front of you. USE BOTH HANDS to carry the microscope. Use one hand to lift the microscope by grasping firmly on the ARM. Put the other hand underneath the BASE to support it.
3. Make sure that the power cord is not dangling before you begin to walk with the microscope. Take it DIRECTLY to your lab table and place it on the counter.
4. Clean the lenses, especially the oculars, before using. Use the special LENS PAPER to clean the lenses. NEVER USE anything else. Paper towels, Kim-wipes, and Kleenex can scratch the lenses. If any liquid gets on the lenses at any time, wipe it off immediately. Don't let it dry!
5. When you are finished using the microscope follow these steps:
   - Remove the slide. Clean and dry the stage.
   - Lower the stage to its lowest level. Center the stage.
   - Clean the lenses. Use lens paper only!
   - Rotate the nosepiece so that the 4× objective is in the view position.
   - Unplug the light cord. Wrap it loosely around the lens head.
   - Return the microscope to its correct place in the cabinet.

THE COMPOUND MICROSCOPE

You should locate the following parts of the microscope. Be able to label them on a diagram (refer to the diagram in the previous exercise). You must also know what the functions of the parts are. Refer to the previous exercise for a description of the parts of the microscope.

1. Base
2. Arm (Body Tube Support)
3. Light source
4. Light switch
5. Rheostat
6. Stage
7. Stage clips or mechanical stage
8. Condenser
9. Condenser control knob
10. Iris diaphragm
11. Body tube
12. Ocular lens (usu. 10×)
13. Revolving nosepiece
14. Objective lenses (scanning, low power, high power, oil immersion)
15. Coarse-focusing knob
16. Fine-focusing knob

Magnification

The magnification (power) of each lens is inscribed on the lens.

The TOTAL MAGNIFICATION is obtained by multiplying the power of the objective by the power of the ocular. THIS MUST BE NOTED ON ALL DRAWINGS.

Resolving Power

The resolving power is the ability of the microscope to distinguish between two objects. Most student microscopes can resolve objects that are 0.5 microns apart. Resolution can be increased by using smaller wavelengths of light to form the image. The best resolution is achieved by using blue wavelengths of light. You can also increase resolution by using a lens with a higher numerical aperture.

Contrast

In order to distinguish different parts of an object, you must be able to obtain sufficient contrast. Sometimes a specimen must be stained in order to make certain objects visible. You can also improve contrast by reducing the amount of light. This may be necessary when viewing an unstained specimen. Alter the amount of light by adjusting the light intensity knob (rheostat) or the iris diaphragm. The light intensity control should normally be set to about 1/3–1/2 the maximum. LEAVE THE CONDENSER IN THE HIGHEST POSITION. Do not adjust the light by raising or lowering the condenser.

Focusing

A microscope is focused by increasing or decreasing the distance between the specimen and the objective lens.

1. Rotate the 4× objective into the viewing position.
2. Use the coarse-focusing knob to raise the stage to its highest position. LOOK FROM THE SIDE (NOT THROUGH THE OCULAR) WHILE YOU DO THIS.

3. While looking through the ocular, SLOWLY lower the stage by turning the coarse-focusing knob away from you until the object comes into focus. Use the fine-focusing knob to bring the object into sharp focus.

4. Once you have the image focused with the 4× objective, you can increase the magnification by switching to the 10× objective. Be sure that the object is centered in the field. Leave the stage in position and rotate the eyepiece until the 10× objective slides into the view position. You will hear a click. You can adjust the focus using the coarse and fine focus knobs.

5. If you need to increase the magnification again, repeat the procedure, and slide the 40× objective into the view position. When you adjust the 40× lens, you should only use the fine focus knob. NEVER USE THE COARSE FOCUS KNOB WITH THE 40× OBJECTIVE, or you may crack the slide or damage the lens.

6. If you are using the oil immersion lens, you will need to place a small drop of immersion oil on the slide BEFORE you rotate the oil immersion lens into position. When you focus using the oil immersion objective, you must use the fine focus only. NEVER USE THE COARSE FOCUS KNOB WITH THE OIL IMMERSION LENS. You must also be careful to make sure that you do not drag the 40× objective through the oil. If you get oil on any other lens or part of the microscope, you should clean it off immediately. Your instructor will demonstrate the proper technique for using the oil immersion lens for those labs in which it will be used. You will not need to use this lens for these exercises.

Notes

If an object is centered and in focus with one objective, it will remain centered and in focus when you switch to another objective. Microscopes that have this quality are said to be parcentric and parfocal. You will probably need to make minor adjustments to the focus however. Use the FINE FOCUS knob.

As magnification increases, the (1) working distance, (2) diameter of field, and (3) light intensity are reduced. You should know this relationship.

Be aware—the direction of movement (i.e., right, left, "up," "down") are reversed while you are looking through the microscope.

Assignments for This Lab Exercise

1. Become familiar with the parts of the microscope. Using the provided slides, practice focusing with the coarse and fine focus knobs. Observe the slides through the 4×, 10×, and 40× objectives. Try adjusting the light intensity with the rheostat. Open and close the iris diaphragm to see how it affects the amount of light and the contrast. Raise and lower the condenser to see the same effects. Remember—if you are using a binocular microscope, you may need to adjust the width of the ocular lenses so that you can see the image clearly with both eyes.

   Examine the following slides and make the required observations. Remember—all drawings must conform to the format which was given to you at the first lab. Remember to label all structures, indicate the objective used, and the total magnification.

2. **Magnification—the Letter "e"**

   Using the prepared slide provided to you, draw and label the letter "e" as you observe it using the 4×, 10×, and 40× objectives. Note the changes that occur as you increase the magnification.

   As you move the slide to the right, in which direction does the image move? _____

   If you move the slide away from you, in which direction does the image move? _____

   What happens to the image as you increase the magnification? _____

   What happens to the light as you increase the magnification? _____

3. **Depth of Field—the Crossed Threads**

   The depth of field is the ability of the microscope to distinguish between layers. Use the slide labeled "Crossed threads." Draw and label the threads as they appear using the 4×, 10×, and 40× objectives. Pay attention to which threads are in the plane of focus on different magnifications (depth of field).

Are you able to see all three threads in focus at one time when you are using the 4×, the 10×, and the 40×? _____.

What is the relationship between the magnification and the depth of field?

4. **Making a Wet Mount—Human Epithelial Cells**

   Make a wet mount of human epithelial cells in the following manner:

   a. Use a flat toothpick to gently scrape some material from the inside of the oral cavity (cheek).

   b. Place a drop of water or methylene blue on a clean slide. Gently swirl the toothpick in the drop to transfer cells to the slide.

   c. Gently apply a coverslip. Try to avoid creating bubbles. It is usually best to lower the cover slip slowly, at an angle until the edge of the coverslip touches the mounting fluid. Let the fluid flow along the edge of the coverslip. Then continue to lower the coverslip until it is flat. See Figure 9.1.

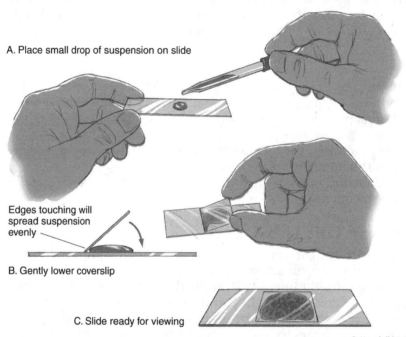

Figure 9.1
Wet Mount

d. Examine under the microscope, using the 4×, 10×, and 40× objectives. Look for some squamous epithelial cells (See Fig. 9.2). Draw and label the cells.

e. Be sure to discard your toothpick and your slides in the biohazard waste container, provided by your instructor. DO NOT place them in the regular trashcan.

5. **Measuring the Diameter of Field**

   Place a transparent plastic ruler on the stage. Using the 4× objective, focus and measure the diameter of the field in millimeters (mm).

   Record your measurement. _____.

   Measure the diameter of field at 100×. _____.

   Measure the diameter of field at 400×. _____.

   Go back to the "letter e" slide and try to estimate the height of the letter in mm. _____.

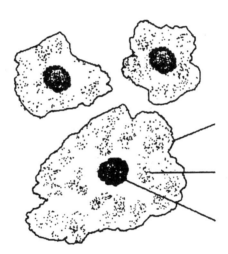

Figure 9.2
Human Epithelial Cells

Name _____ Section _____ Date _____

Exercise 10   Plant Cells

PRINCIPLE

The purpose of this exercise is to give the student an introduction to the general structure of plant cells, and to allow the student to recognize some of the major organelles contained in plant cells that can be detected at the level of light microscopy.

OBJECTIVES

After completing this exercise, you should be able to:

1. Locate and draw the following parts of a plant cell
   - nucleus
   - nucleolus
   - nuclear membrane
   - cytoplasm
   - vacuole
   - chloroplast
   - cell wall
   - cell membrane
   - stoma
   - amyloplasts
2. Be able to label these structures on a diagram
3. Recognize the following types of cells:
   - epidermal cells—thick walled, regular shape, protective functions
   - parenchymal cells—thin walled, irregular, various functions
   - guard cells—paired, on the underside of leaves, gas and water exchange

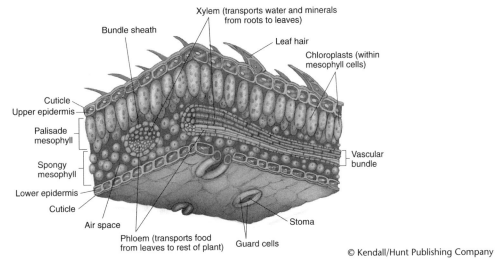

Figure 10.1

55

ASSIGNMENTS FOR THIS LAB EXERCISE

Elodea Leaf

Elodea is a common water plant. Its cells are typical of highly photosynthetic plants and contain many chloroplasts.

1. Make a wet mount: Using forceps, remove a young leaf from the tip of a plant shoot.
2. Put a drop of water on a glass slide and drop the leaf into it.
3. Gently apply a coverslip. Examine using the 4×, 10×, and 40× objectives.
4. Draw ONE view. Label as many parts as you can find.

Zebrina

1. Repeat the same procedure with a *Zebrina* leaf.

   You may need to peel a section from the leaf to make the wet mount. Mount the leaf with the bottom (purple) side facing up.
2. Examine using the 4×, 10× and 40× objectives
3. Find, draw, and label a pair of GUARD CELLS. Be sure to include the stoma and the surrounding epidermal cells in your drawing. *Note:* The guard cells will be green and the epidermal cells will be purple.

Tomato

1. Use an eyedropper to pick up a sample of tissue from the pulpy flesh.
2. Place a drop of the sample on a glass slide.
3. Gently apply a coverslip. Examine using the 4×, 10×, and 40× objectives.
4. Find some thin-walled parenchymal cells. Draw and label as many parts as you can find. (Draw one magnification only.)

Lettuce (optional)

1. Slice a section and make a wet mount as described for *Elodea*.
2. Examine using the 4×, 10×, and 40× objectives.
3. Find, draw, and label a pair of GUARD CELLS. Be sure to include the surrounding epidermal cells in your drawing.

Potato Starch Cells

1. Use a scalpel or sharp single-edge razor to slice through the tissue.
2. Try to cut sections as thin as possible. Use a smooth, continuous slicing method and as much of the blade as possible. DO NOT put a lot of pressure on the tissue. DO NOT saw or chop.
3. BE CAREFUL. DO NOT CUT YOUR OWN FINGERS!!
4. Place the section of potato on a glass slide. Add a drop of starch solution.
5. Gently apply a coverslip. Examine using the 4×, 10×, and 40× objectives.
6. Draw ONE view. Label as many parts as you can find. Be sure to include the hilus, amyloplasts, and striations in your drawing.

Onion Epidermal Cells

1. Break a piece of onion scale (ring).
2. Use forceps to strip off a small piece of the epidermis from the inside surface.
3. Place on a glass slide in a drop of water or methylene blue.
4. Flatten and smooth the section. Apply a coverslip.
5. Examine using the 4×, 10×, and 40× objectives.
6. Draw and label an epidermal cell.

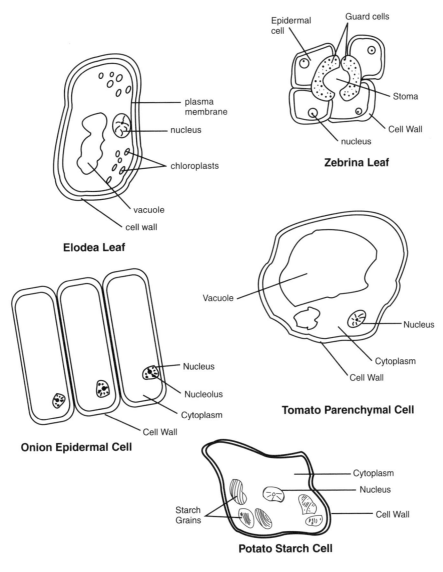

Figure 10.2
Plant Cells

Exercise 10 Plant Cells 57

Name _____ Section _____ Date _____

Exercise 11 Animal Cells

PRINCIPLE

The purpose of this exercise is to introduce the student to the general structure of animal cells. A variety of cells are examined to demonstrate particular organelles or structural specializations. A series of embryonic cells is examined to introduce the student to the concept of developmental stages.

OBJECTIVES

After completing this exercise, you should be able to:

1. Locate and draw the following parts of an animal cell.

    All Cells
    - Plasma membrane
    - Nucleus
    - Cytoplasm

    Specialized Structures
    - Cilia and brush border (striated border on ciliated epithelial cells)
    - Sarcomeres (striations of skeletal muscle cells)
    - Haversian system of bone, including: osteocyte, lamellae, lacunae, Haversian canal, canaliculi, blood vessels
    - Sperm—head and tail
    - Graafian follicle—oocyte, antrum, ovarian tissue
    - Neuron—dendrites, cell body, axon
    - Cartilage—lacuna, chondrocyte
    - Embryo—stages, blastocoel, endoderm, ectoderm, mesoderm, archenteron

2. Be able to describe the structure and its function. Be able to locate and label them on a diagram.
3. Recognize the following types of cells:
    - Blood cells
        A. Erythrocytes (red blood cells)
        B. White blood cells—granulocytic series
            1. *Neutrophils*—neutral staining granules
            2. *Eosinophils*—red staining granules
            3. *Basophils*—dark staining granules
        C. Non-granulocytic series
            1. *Lymphocytes*—small dark blue round nucleus
            2. *Monocytes*—large, pale "foamy" cytoplasm
        D. Thrombocytes (platelets)—very small, dark blue
    - Epithelial cells: Squamous and Columnar
    - Goblet Cells
    - Three Types of Muscle

- Bone (Haversian systems)
- Reproductive cells: Sperm; Graafian Follicle (ovary)
- Developmental stages of the Sea Star

    | fertilized egg | 4-cell stage | morula |
    | blastula | blastocoel | gastrula |

- Neuron
- Cartilage

Assignments for This Exercise

All slides used in this exercise are already prepared, fixed, and stained. Use the drawings and photos in your atlas as a reference. Your instructor may also provide reference charts in class.

1. **Blood (Human or Frog)**—Your instructor will assign you to draw a variety of blood cells from the following types:

    a. Erythrocyte
    b. Thrombocyte
    c. Neutrophil
    d. Eosinophil
    e. Basophil
    f. Lymphocyte
    g. Monocyte

2. **Intestinal Epithelium (Frog)**

    Draw a section of the intestinal epithelium which includes columnar epithelial cells, a striated border, and goblet cells. Label all structures found. THIS SHOULD BE DONE AS A SINGLE DRAWING.

3. **Three Types of Muscle**

    Draw and label a small section of skeletal muscle. This will include several cells. Be sure to indicate several nuclei and the striations.

    Draw and label a small section of smooth muscle. Note that the cells are spindle shaped, and have only one nucleus. The cells lack the striations found in skeletal muscle, and therefore appear smooth.

    Draw and label a small section of cardiac muscle. Note that the cells have one nucleus and obvious striations. The cells have a branching structure. Special junctions between the cells called intercalated discs will appear as small, dark, vertical lines. You may have to go in and out of focus with your fine adjustment to see these lines. These junctions allow cardiac muscle to contract as a single unit.

4. **Generalized Animal Cell**

    Draw and label a squamous epithelial cell.

5. **Human Bone (Cross Section of Ground Bone)**

    Draw and label at least one Haversian system. Indicate the following structures on your drawing: Haversian canal, blood vessel, lamellae, lacunae, osteocytes, and canaliculi.

6. **Sperm (Human or Bull)**

    Draw several sperm cells. Label the head and tail of at least one cell.

7. **Graafian Follicle (Ovary)**

    Within the ovary, locate and draw a Graafian follicle. This will generally be the largest follicle in your section. Label the following: ovarian tissue, oocyte, and antrum.

8. **Neuron**

   Draw and label a portion of a neuron that includes: the cell body, dendrites, and the axon.

9. **Hyaline Cartilage (Human)**

   Draw a section of cartilage that contains a few cells. Label the lacunae and chondrocytes.

10. **Starfish Development** (*Asterias*—the sea star)

    Draw and label the stages of development from fertilized egg to gastrula. Your drawing should include at least 6 stages:

    fertilized egg, 4-cell stage, morula, early blastula, late blastula with beginning invagination, and gastrula

    You should indicate the following features of development:

    the multiplication of cells

    the development of the blastocoel and the archenteron

    the process of invagination

    gastrulation and the development of the three tissue layers: endoderm, ectoderm, and mesoderm

Name _____ Section _____ Date _____

Exercise 12: Diffusion: Osmosis and Dialysis

PURPOSE

To observe the properties of diffusion under varying conditions.

PROCEDURES

1. Simple diffusion
   a. Fill a 100 ml beaker with tap H_2O, place beaker on white background. Gently lay 1–2 drops of India ink on top of the water. Observe and record results at 10-minute intervals for at least 60 minutes.
   b. Fill a 100 ml graduated cylinder with 75, ml tap H_2O. Insert a hollow glass tube to the bottom of the cylinder and pour a small amount of K-permanganate crystals through the hollow tube. *Gently* remove the glass tube. Observe the cylinder over at least 48 hours. Take care to avoid mixing the contents of the cylinder.

2. Osmosis in an artificial cell system
   a. Take 2 lengths of dialysis tubing; open the tubing and *securely* tie one end with the string supplied. Fill each tube with Karo syrup and secure the top of the bag with string. All bags should be about the same size. Check for leaks.
   b. Weigh each bag and record results.
   c. Place 1 bag in a dish or beaker filled with tap H_2O.
   d. Place 1 bag in a dish or beaker filled with distilled H_2O. Be sure bags are submerged.
   e. Weigh bags at 15-minute intervals for *at least* 60 minutes. Record results.

3. Osmosis in living cells
   a. Tissue
      1. Cut 3 small cubes of potato of approximately equal size.
      2. Measure size of each cube with a ruler and record.
      3. Place 1 cube in a watch glass containing the following solutions:

         tap H_2O 10% salt   distilled H_2O

      4. Gently feel each cube and measure each cube at 10-minute intervals for at least 60 minutes. Record results.
   b. Cells
      1. Place fresh *Elodea* leaf in tap H_2O on a clean glass slide. Draw, label cells at 43× magnification, and briefly describe their appearance.
      2. Transfer leaf to 10% salt solution; observe cells at 43×. Draw, label, and describe their appearance.
      3. Remove leaf from slide, quickly dip 2–3 times in beaker of *tap* H_2O; transfer leaf to clean slide with distilled H_2O. Observe cells at 43× magnification. Draw, label, and describe their appearance.

Name _____ Section _____ Date _____

RESULTS

1. *Simple Diffusion*

Time	Solid	Liquid	Gas
0			

Name _____ Section _____ Date _____

RESULTS

2. *Osmosis in Living Cells—Tissue (Potato)*

Time	Tap H$_2$O	Distilled H$_2$O	Salt H$_2$O
0			

Name _____ Section _____ Date _____

Part IV

Metabolism

Name _____ Section _____ Date _____

Exercise 13, Part 1: Photosynthesis

OBJECTIVES

After completing this exercise, you should be able to:

1. Describe the arrangement of tissues within a leaf and its relationship to light absorption and photosynthesis.
2. Write summary equation for photosynthesis.
3. Describe the effects of light intensity and other variables on the rate of photosynthesis.
4. Describe the relationship between photosynthesis and cellular respiration.

All of life is dependent upon its environment. This is especially true when it comes to how organisms obtain the **energy** needed to stay alive as well as to move, grow, reproduce, and do the other things you associate with life. In the process of **photosynthesis,** plants (and other autotrophs) use the energy of sunlight to synthesize carbohydrates that serve as the major energy molecule. The leaf is the usual site of photosynthesis. After transport to cells, the energy of the carbohydrates is "harvested" within the cytoplasm of the cell and the mitochondria in the process of **cellular respiration.** The mitochondria in both plants and animals function in cellular respiration. This process will be studied in Exercise 15.

In this investigation, we begin first with a look at the internal anatomy of a typical leaf. A leaf is a highly efficient solar collector. The numerous cells within a leaf are embedded with chloroplasts, each with thousands of molecules of chlorophyll. While reflecting green light, chlorophyll absorbs most of the other portions of the color spectrum, making this energy available for driving the initial steps of photosynthesis (light-dependent reactions). Next, you will measure the rate at which photosynthesis occurs under various environmental conditions by counting oxygen bubbles.

Activity 1: Internal Anatomy of a Leaf

Examine a prepared slide of a cross section through a leaf of lilac *(Syringa)* or other such leaf. Locate the following structures on the slide as you label Figure 13.1: **veins** composed of **xylem** (large thick-walled cells) and **phloem** (smaller cells); **mesophyll** regions composed of a tightly packed **palisade** and loosely arranged **spongy** portions; **upper epidermis** and **lower epidermis,** the latter with **guard cells,** each pair surrounding an opening known as a **stoma** (pl. stomata); and a thin, waxy, noncellular **cuticle,** which covers the epidermis. The arrangement of these tissues within a leaf is primarily related to the process of photosynthesis that is accomplished within the numerous chloroplasts of the mesophyll cells.

PHOTOSYNTHESIS

Photosynthesis occurs in two separate sets of reactions: (1) **light-dependent** followed by (2) **light-independent.** In the former, light is absorbed, resulting in the formation of molecules needed for the latter. The completion of the light-independent reactions produces a sugar called PGAL, a 3-carbon compound. Molecules of PGAL quickly pair to form glucose, a 6-carbon sugar from which other carbohydrates may be produced.

From *Investigations in Biology* 4th edition by Marion Wells et al. Copyright © 2001 by Kendall/Hunt Publishing Company. Reprinted by permission.

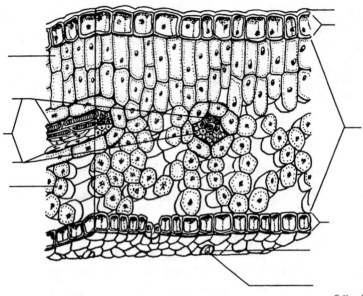

Figure 13.1
Cells and Tissues of a Leaf

Activity 2: Light Intensity and Photosynthesis

For present purposes, you will use a simplified chemical equation to represent photosynthesis:

$$6CO_2 + 6H_2O \xrightarrow[\text{light}]{\text{chlorophyll}} C_6H_{12}O_6 + 6O_2$$
$$\text{carbon dioxide} \quad \text{water} \qquad\qquad \text{glucose} \quad \text{oxygen}$$

You will see in Exercise 15 that aerobic respiration is the reverse of the process above.

The rate of any chemical reaction may be measured as either the amount of product produced or raw materials consumed per unit time. Thus, (referring to the equation above) the rate of photosynthesis can be determined by measuring the formation of glucose or oxygen, or the disappearance of carbon dioxide or water. For practical considerations, in this activity, we will be using oxygen production as an indicator of photosynthetic rate. Any actively photosynthesizing plant will release oxygen; if the plant is underwater, bubbles of oxygen are easily visible and can be counted. In this investigation you will be utilizing waterweed, *Elodea* sp.

Materials

Equipment	Reagents/Specimens
flood lamp	water
1000 ml beaker	*Elodea* sprig
test tubes	baking soda
test tube rack	
razor blade	
red cellophane	

Procedure

Work in groups as directed by your instructor.

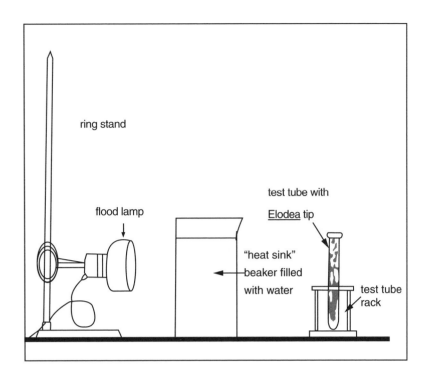

Figure 13.2
Oxygen Production in *Elodea*

Control Group

1. Using a clean, sharp razor blade, cut at an angle a 10 cm length of *Elodea* stem tip while the plant is underwater.
2. Immediately place the *Elodea* tip upside down in a test tube so that the cut end is about 3 cm from the top of the tube.
3. Fill the tube to the top with tap water.
4. As shown in Figure 13.2, position the *Elodea* tube 20 cm from the flood lamp. (The beaker of water acts as a "heat sink" absorbing heat from the flood lamp and allowing the study of the effects of light rather than heat.)
5. Allow 5 minutes for the plant to become stabilized, then count the number of bubbles released from the cut end of the *Elodea* in each of 3 consecutive 3-minute periods. Record in Table 13.1.

Experimental Groups

Light Intensity

1. Move the *Elodea* to a point 30 cm from the lamp and repeat Step 5 above.
2. Move the *Elodea* to a point 40 cm from the lamp and repeat Step 5 above.

Carbon Dioxide Concentration

Follow the same procedure as with the control group, with the exception that after Step 4, add a pinch of baking soda to the *Elodea* tube.

Temperature

Follow the same procedure as with the control group except remove the heat sink after Step 4.

Wavelength of Light

Follow the same procedure as with the control group except place a frame of red cellophane between the heat sink and the flood lamp.

Table 13.1
Response of *Elodea* to Light, Increased Carbon Dioxide, Increased Temperature, and Change in Wavelength of Light

Control/Experimental Tubes	Number of Oxygen Bubbles			Mean ($\bar{X}$)	Standard Error	± 2 Standard Errors
	1st 3 min.	2nd 3 min.	3rd 3 min.			
Control Tube (20 cm)						
Light Intensity (30 cm)						
Light Intensity (40 cm)						
Control Tube (20 cm)						
CO_2 Concentration (baking soda added)						
Control Tube (20 cm)						
Temperature (without heat sink)						
Control Tube (20 cm)						
Wavelength of Light (with red cellophane)						

Name _____ Section _____ Date _____

Exercise 13, Part 2 | Photosynthesis Laboratory Report

1. Indicate the functions performed by these parts of a leaf:
   a. Cuticle _____
   _____
   _____

   b. Veins _____
   _____
   _____

   c. Stomata _____
   _____
   _____

   d. Guard cells _____
   _____
   _____

   e. Mesophyll cells _____
   _____
   _____

2. In the photosynthesis experiments:
   a. What is the dependent variable? _____
   b. Identify the independent variables involved in this investigation. _____
   _____
   c. What role is played by the sodium bicarbonate? _____
   _____
   _____
   d. Why is the beaker of water necessary? _____
   _____
   _____
   e. State in a paragraph the conclusions you reached from this investigation.
   _____
   _____
   _____
   _____

Name _____ Section _____ Date _____

Exercise 14: Separation of Leaf Pigments by Paper Chromatography

Leaves contain a variety of pigments, such as chlorophyll a (blue-green), chlorophyll b (yellow-green), carotene (orange), and xanthophylls (yellow). Because the pigments have different molecular weights, they move at different rates up a piece of absorbent paper when they are in solution. This makes it possible to separate these pigments by a process known as paper chromatography.

MATERIALS

- Pigment extract from spinach leaves
- Strips of chromatography paper
- Scissors
- Large test tubes with corks and pins
- Test tube rack
- Pipettes
- Chromatographic solvent (8% acetone; 92% petroleum ether)
- Container labeled "waste chromatographic solvent"

METHODS

1. Obtain a strip of chromatography paper that is approximately the same length as the large test tube that you will be using. Handle only the edges of the paper!
2. Trim one end of the paper as shown in Figure 14.1. (Be sure to discard the bits of paper that you trim away into the trash container!)

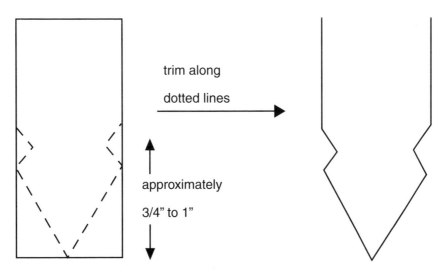

Figure 14.1

3. Using a small-tipped pipette, place ONE small drop of the pigment extract from spinach leaves (already prepared) onto the space between the two notches on your strip of paper. Make the drop as small as possible!

4. Allow the drop on the paper to dry completely, then add another small drop to the same place between the notches. Allow the second drop to dry completely.

5. Continue to add one small drop at a time until a total of 10 drops (or more, as per instructor) has been applied to the same spot. Be sure to allow the spot to dry completely after each application to concentrate the pigment in one spot on the paper!

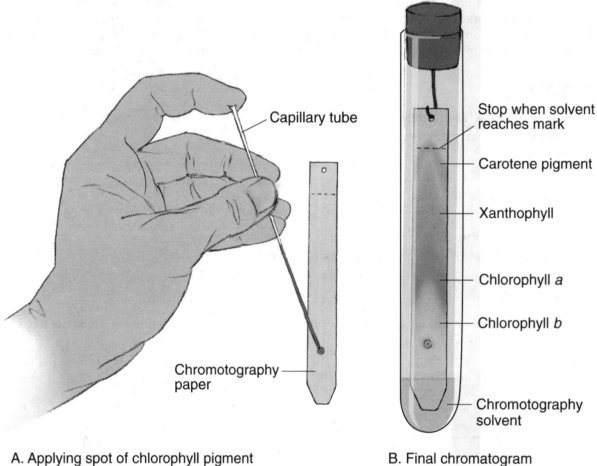

A. Applying spot of chlorophyll pigment   B. Final chromatogram

Figure 14.2
Paper Chromatography

6. Suspend the strip of chromatography paper onto the bottom of a cork by a hook or pin. Place it in a dry, clean test tube to check the length of the paper. The paper should almost touch the bottom of the test tube. Adjust the length as necessary by moving it up or down the pin. (Cut a little paper off the top if it is still too long.)

7. Remove the strip of paper together with the cork from the test tube. Add a small amount of chromatographic solvent to the test tube to fill the bottom, curved portion of the test tube.

8. Reinsert the paper and cork tightly into the test tube, making sure that the *tip* of the paper is in the solvent. Place the test tube upright (as straight as possible!) into a test tube rack.

9. When the solvent has moved almost to the top of the paper strip (about 30 minutes), remove the paper from the test tube and allow it to dry.

10. Attach the strip of chromatography paper to a plain white piece of paper. Carefully label the bands that you see by indicating the name of each pigment and its color.

You should see five bands of color, representing different pigments that have separated from the mixture of pigments, in the following order:

Carotene (yellow-orange to orange)—has the lowest molecular weight and moves the most rapidly up the paper, so it should be near the top of the paper

Tannin (gray) - spinach

Xanthophylls (yellow) (yellow to brown)

Chlorophyll a (blue-green or dark green)

Chlorophyll b (yellow-green or light green)—has the highest molecular weight and moves the slowest up the paper, so it will be nearest to the bottom of the paper

11. Clean up: Simply decant the remaining solvent from the test tube into the container labeled "waste chromatographic solvent" and place the test tube upside down in the test tube rack. Do *not* rinse the test tubes with water!

Make sure your lab group does not discard the unused pigment extract; it will be collected by your instructor!

Please Note: Information regarding chromatography can be found on reserve in the Learning Resource Center.

Name _____ Section _____ Date _____

Exercise 15 Fermentation Lab

OBJECTIVES

After completing this exercise, you should be able to:

1. Describe alcohol fermentation and name and describe environmental factors that influence rates of fermentation.
2. Propose hypotheses about factors that influence fermentation and make predictions based on these hypotheses.

PROCEDURE

Alcohol Fermentation

For this lab we will use yeast, which is a single-celled fungus. Yeast cells are capable of using glucose for growth through a process called fermentation. The yeast produce both alcohol and carbon dioxide (CO_2) as a result of this fermentation. In this lab, the yeast will be provided Karo syrup, which provides both glucose and fructose as a substrate for growth. The rate of fermentation will be determined by the production of CO_2.

Figure 15.1 shows the setup each team will use to collect the CO_2 produced by the yeast. Fermentation takes place in the tube on the right (fermentation tube), which contains the yeast. The fermentation tube is capped with a rubber stopper and plastic tubing to collect any CO_2 produced by the yeast. The plastic tubing leads from the fermentation tube to a collection tube, which is oriented upside down (and contains water). This tube is designed to collect CO_2 produced by the yeast during fermentation. The CO_2 that is produced during fermentation is collected in the collection tube and displaces the water inside of the tube. The amount of water displaced is proportional to the amount of fermentation. The amount of water displaced will be measured in millimeters (mm) as a measure of the amount of CO_2 produced.

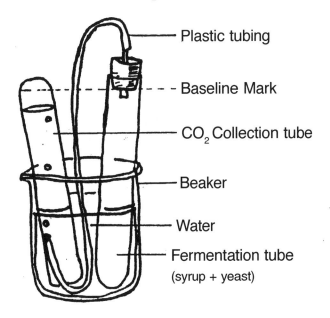

Figure 15.1

LAB SETUP FOR MEASURING CO_2 PRODUCTION IN ALCOHOL FERMENTATION

1. Each lab station has three fermentation setups containing six large test tubes, three pieces of plastic tubing that have been inserted into rubber stoppers, and three beakers (400 or 600 ml). Use a wax pencil and label three of the test tubes 1, 2, and 3 and set them aside. Assemble the setups one at a time following the steps below.
2. Fill the sink with **hot** water.
3. Insert the end of the plastic tubing into one of the test tubes. This tube will be the CO_2 collection tube. Submerge the collection tube and plastic tubing in the sink of hot water.
4. Submerge the beaker in the sink of hot water. Place the collection tube in the beaker in an inverted position.
5. Bring the beaker out of the water. One end of the plastic tubing should still be inserted in the collection tube. Hold up the other end of the tubing (the one with the rubber stopper on it) so that the water will not be siphoned out.
6. Pour some of the water out of the beaker so that the water level is about 1 cm from the top of the beaker.
7. Check the tubing for kinks. If there are kinks in the tube, CO_2 cannot get through the tubing and you will have to set up the apparatus again.
8. Assemble all three of the setups before adding the fermentation mixture.
9. Make up the fermentation solutions for tubes 1, 2, and 3 according to Table 15.1.

Table 15.1
Contents of the Fermentation Solution

	Fermentation Tube (volume in ml)		
	Tube 1	Tube 2	Tube 3
Water	4	2	0
Yeast suspension	0	2	4
Syrup	3	3	3

10. Swirl each test tube gently to mix the water, yeast, and syrup. Place one test tube in each beaker.
11. Put the rubber stoppers in the fermentation vials. This will force most of the water out of the tubing.
12. After the air bubbles have cleared the tubing (about 1 minute) mark the water level on each collection tube with a wax pencil. This makes the baseline for the experiment.
13. At 5-minute intervals measure (in mm) the distance from the baseline mark to the water level. Continue taking data for 40 minutes. Record your data in Table 15.2.

HYPOTHESIS STATEMENT

In the space below, write a hypothesis statement for this experiment. Your hypothesis should relate to the ingredients of your fermentation solution.

RESULTS

1. Complete Table 15.2.

Table 15.2

Tube	0	5	10	15	20	25	30	35	40
1									
2									
3									

Time (minutes)

2. Construct a graph to illustrate your results.
    a. What is the independent variable? Which is the appropriate axis for this variable?
    b. What is the dependent variable? Which is the appropriate axis for this variable?
    c. Choose an appropriate scale and label the X- and the Y-axis. Make a title for your graph.

DISCUSSION

1. Which test tube had the highest rates of fermentation? Explain why.

2. Which test tube had the lowest rates of fermentation? Explain why.

3. Which fermentation tube was the control?

4. Why were different amounts of water added to each fermentation solution?

5. Was the hypothesis supported by the results? Use your data to support your answer.

6. What are some other independent variables that could affect fermentation rates?

Name _____ Section _____ Date _____

Exercise 16: Cell Respiration— Go for the Burn!!

INTRODUCTION

We know that all cells need energy to live and function. We also know that our cells, animal cells, get their energy from food. Metabolism is the name for all of the chemical processes that occur in our cells that enable our cells to utilize the energy stored in food. What happens when our cells are working too hard? Today we will find out.

METHODS

1. Work in pairs. One member of each team will put his/her hand on the table with the palm facing UP. Then he/she will open and close his/her hand as rapidly as possible. MAKE SURE THAT EACH TIME THE HAND IS OPENED THE FINGERS ARE COMPLETELY FLATTENED AGAINST THE TABLE.
2. The other member of the team will count the number of times the hand is closed in 20-second trials. After each 20-second interval, quickly record the number and immediately start the next trial. You will do 10 trials, each 20 seconds long.
3. Enter your results in the data table you create below. (USE A RULER!)
4. Graph your results on the back. The trial number will go on the X-axis and the number of times the hand was closed goes on the Y-axis.

RESULTS

Data Table 1: Number of times a hand opens and closes in 20-second time intervals

Graph 1: Number of times a hand opens and closes in 20-second time intervals

DISCUSSION QUESTIONS

(Answer the following questions)

1. What did the muscle feel like after repeated trials?

2. Use your knowledge and notes from Cellular Respiration to explain what caused this sensation. What was happening in your muscles? (Hint: your answer should go beyond "My muscle cells were tired.")

3. Summarize your results (e.g., did your hand close slower, faster).

4. Write a short paragraph that relates this experiment to the idea of energy. Use words such as ATP (cellular energy), glucose, metabolism, fatigue, and lactic acid in your paragraph.

Part V

Cell Division

Exercise 17: Cell Division

FACTS AND INFORMATION

All cells arise from preexisting cells. In eukaryotes, we call the time frame between cell division the **cell cycle.** The cell cycle is generally divided into two phases, interphase and cell division. Figure 17.1 illustrates a typical eukaryotic cell cycle.

As can be seen, the cell spends most of its life in interphase. **Interphase** is divided into three stages. The first part of interphase is called G_1 and is characterized by a period of growth for the cell. The next stage is called S during which DNA replication in the nucleus takes place. During S, the DNA of a chromosome is replicated resulting in the formation of two sister chromatids (Figure 17.2).

During S, proteins associated with DNA replication are also synthesized. Although represented in Figure 17.2 as condensed bits of DNA, the chromosomes throughout interphase are in a diffuse state and the genetic material is referred to as chromatin. Following S, the cell moves into the last stage of interphase called G_2. This stage is characterized by growth of the cell and replication of some cellular organelles in preparation for cell division.

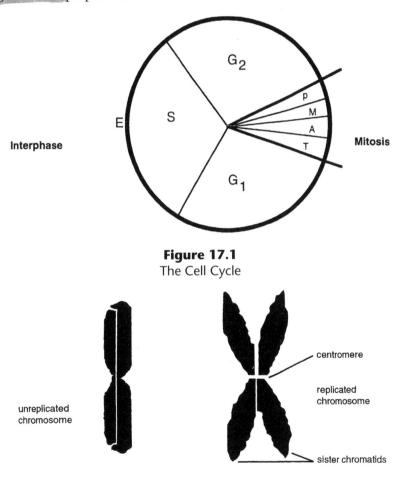

Figure 17.1
The Cell Cycle

Figure 17.2
Chromosome Before and After Replication

From *Introductory Biology 103, Laboratory Manual,* 3rd Edition by P. Shields and H. B. Cressey. Copyright © 2002 by Kendall/Hunt Publishing Company. Reprinted by permission.

Cell division consists of two events, the division of the nucleus, **mitosis**, and the division of the cytoplasm, **cytokinesis.** *Mitosis can be defined as the process by which one cell divides into two new genetically identical cells.* For convenience, mitosis is often divided into four stages: **prophase, metaphase, anaphase,** and **telophase.** It is important to realize that these divisions are arbitrary; mitosis is a continuous process. Figure 17.3 and 17.4 illustrate mitosis in both plant and animal cells. It may be helpful to refer to these figures while reading the descriptions of the stages of mitosis.

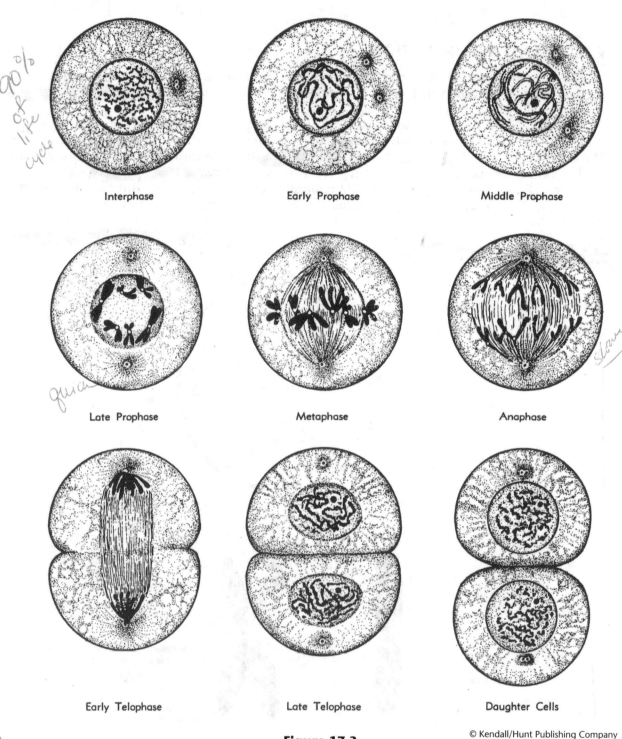

Figure 17.3
Mitosis in Animal Cells

92 Exploring Biology in the Lab

onion root tip

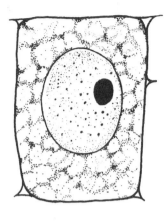
1. Interphase

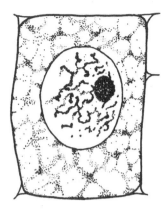
2. Early Prophase

3. Middle Prophase

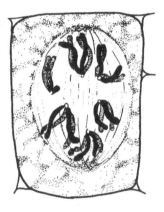
4. Late Prophase

5. Metaphase

6. Anaphase

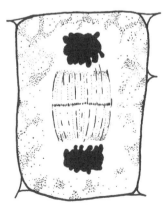
7. Early Telophase

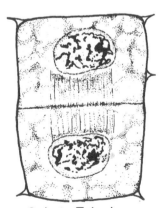
8. Late Telophase

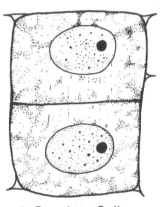
9. Daughter Cells

Figure 17.4
Mitosis in Plant Cells

PROPHASE
- Nuclear envelope disappears
- Nucleolus disappears
- Chromatin begins to condense into chromosomes
- Centrioles begin to migrate to opposite ends of the cell
- Spindle formation begins

METAPHASE
Sister chromatids of each chromosome line up on equator of spindle, attached to the spindle fibers by their centromeres

ANAPHASE
- Centromere divides
- The sister chromatids (now chromosomes) move to opposite ends of the cell

TELOPHASE
- Nuclear envelope reappears
- Chromosomes begin to uncoil and spread out as chromatin
- Spindle disappears
- Cytokinesis begins, evidenced by appearance of cleavage furrow or cell plate

The division of the cytoplasm, cytokinesis, happens as the cell is pinched into two relatively equal portions. The formation of the cleavage furrow signals its beginning in animal cell division. In plants, a cell plate forms in between the new daughter nuclei and will aid in the partitioning of the plant cells.

Thus, at the end of mitosis and cytokinesis, we end up with two new daughter cells, each genetically identical to the other. Mitosis is the mechanism by which all of our somatic cells grow and divide. Somatic cells are those cells which are not sex cells or germ line cells. In humans, eggs and sperm are the cells associated with sexual reproduction, and these cells are produced by the mechanism of meiosis. Before describing the process of meiosis, it is important that we discuss the concept of chromosome number.

All organisms have a chromosome number that is characteristic for that species. For example, humans have 46 chromosomes, while 26 chromosomes characterize frogs, and man's best friend, the dog, has 78. All of these organisms are considered to be diploid. Diploid cells contain two copies of each chromosome, one obtained from the male parent, the other obtained from the female parent. These two copies of the same chromosomes are called homologous chromosomes, or homologues. They are similar in size and shape, but are not genetically identical. Therefore, human sex cells, or gametes, are considered to be haploid, meaning that they contain only one copy of each chromosome.

As can be seen, at sometime in the formation of sex cells, a type of cell division is needed to reduce the number of chromosomes from the diploid state (2n) to the haploid state (n). This special process of cellular division is called **meiosis.** *Meiosis is defined as the form of cellular division that converts a single diploid cell into four haploid cells.* A cell undergoing meiosis will begin by the replication of its genetic material, followed by two successive divisions. The first division, called meiosis I, results in the reduction of chromatids. As with mitosis, each division of meiosis is subdivided into stages. Meiosis is illustrated in Figure 17.5. Follow the illustration as we describe the process.

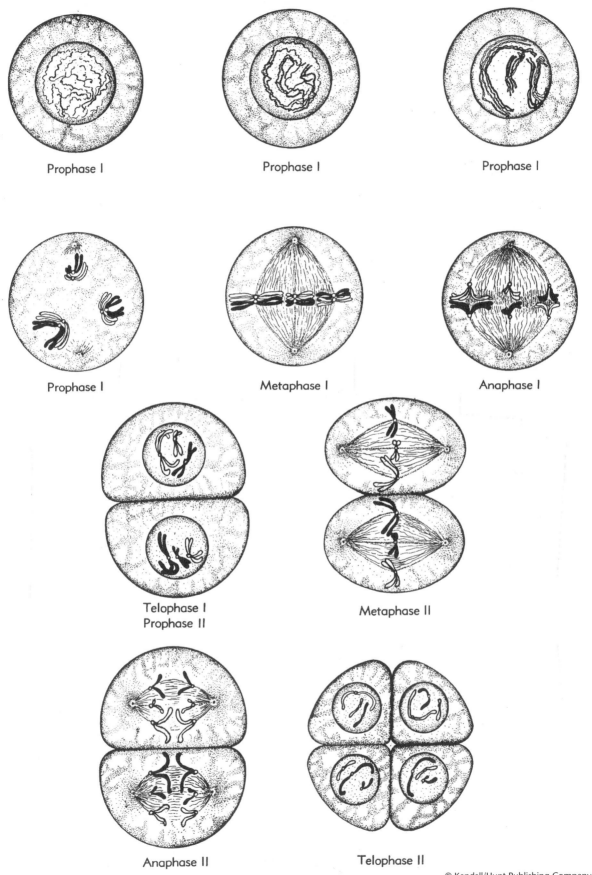

Figure 17.5
Meiosis, Diagram Showing Reduction

MEIOSIS I (Italics indicate differences from mitosis)

Prophase I
- Nuclear envelope disappears
- Nucleolus disappears
- Chromatin begins to condense into chromosomes
- Centrioles begin to migrate to opposite ends of the cell
- Spindle formation begins
- *Homologous chromosomes pair up, forming tetrads*
- *At places where homologues touch, crossing over may occur*

Metaphase I
- *Homologous pairs of chromosomes line up on equator of spindle, each homologue attached to the spindle fibers by their centromeres*

Anaphase I
- *Centromeres do not divide*
- *Homologous pairs separate from each other, each homologue moving to opposite end of the cell*

Telophase I
- Nuclear envelope may or may not reappear
- Chromosomes may begin to uncoil and spread out as chromatin
- Spindle disappears
- Cytokinesis may or may not occur
- Cells are not haploid, with double-stranded chromosomes

There may be a short period before meiosis II begins, and then the cell(s) goes directly into the second division, without any replication of genetic material. The cells at this point are haploid, with replicated chromosomes already. Meiosis II is similar to mitosis, and results in the formation of four haploid cells with single-stranded chromosomes.

Mitosis and meiosis are different in several ways. Only diploid cells can undergo meiosis, while both diploid and haploid cells can undergo mitosis. Mitosis involves only one division and the two new daughter cells are genetically identical to each other and to the parent cell. Meiosis involves two divisions and results in the formation of four daughter cells, all with the possibility of genetic variation.

EXPERIMENTAL PROCEDURES

The Plant Cell
1. Obtain an onion root tip slide.
2. Examine the slide under low power. Can you identify the cell's nucleus, chromosomes, and nuclear membrane?
3. Move to the high power objective on your microscope. Can you identify the parts of the cell now?

4. Move your slide around and find examples of:

   Interphase
   Prophase
   Metaphase
   Anaphase
   Early Telophase
   Late Telophase

5. Draw a sketch of each phase you identified. Label as many parts of the cell as you can.

Animal Cell

1. Obtain the prepared slide of whitefish blastula cells.
2. Repeat part A using the whitefish blastula. When drawing the various phases of mitosis, be sure to point out any differences between this animal cell and the previously viewed plant cell.
3. Draw a sketch of each phase you identified. Label as many parts of the cell as you can.

Modeling Meiosis

1. Work in groups of 3 or 4 for this part.
2. Use the modeling kits available, or obtain pipe cleaners and paperclips from the front table. You will need 2 of each color, for a total of 12 pipe cleaners.
3. Each pipe cleaner represents a chromosome. Replicated chromosomes are two identical pipe cleaners wrapped around each other to form an X shape. A paperclip at the point the two pipe cleaners are joined represents the centromere.
4. With your partners, use the pipe cleaners to diagram what happens in meiosis to a cell whose diploid number is 4. To do this, one dark pink pipe cleaner and one light pink pipe cleaner represent one homologous pair, while the light and dark blue pipe cleaner represent another homologous pair of chromosomes.
5. Record the possible alignments at metaphase I on your datasheet.
6. When you have completely worked out the entire meiotic cycle, and ended with 4 haploid cells, let the instructor know that you are ready to demonstrate the movements of the chromosome through meiosis.
7. Make sure that *all members* of the group can identify the following terms:

Sister chromatid	Haploid
Homologous chromosome	Metaphase plate
Crossing over	Spindle
All stages of meiosis I and II	Synapsis
Diploid	

Whitefish Blastula — Onion Root Tip

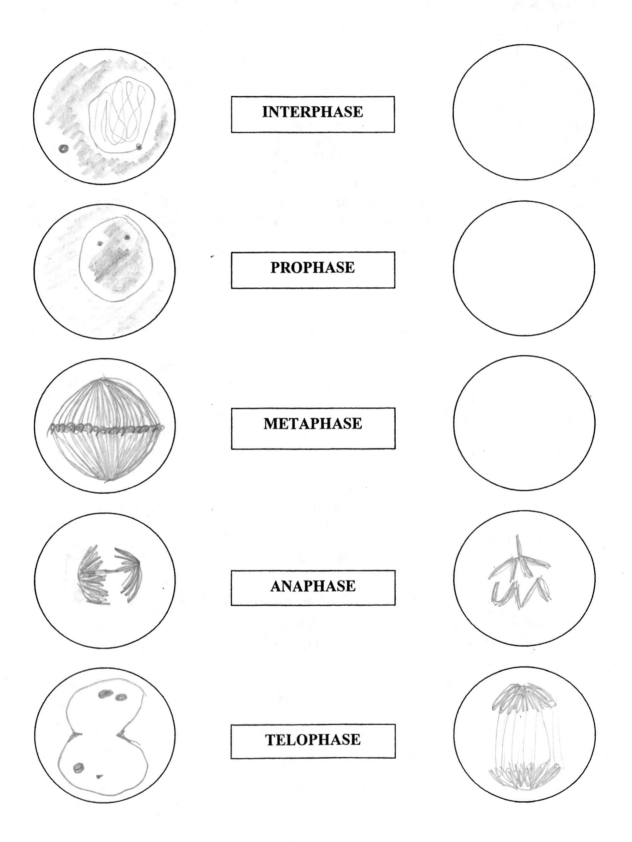

CLEANUP

1. Observe proper microscope handling procedures when putting microscope back into the cabinet:

   Light is turned off

   Lowest power objective is in place

   Any spills have been cleaned with appropriate equipment

   The objectives are as far away from the stage as possible

   The cord is very loosely wrapped around the body

2. Place prepared slides back in original containers.
3. Return pipe cleaners or models to boxes in front.
4. Place stools under laboratory table.

Part VI

Genetics

Name _____ Section _____ Date _____

Exercise 18, Part 1 — Student Worksheet
DNA: The Genetic Code

Nucleic acids are long, chainlike molecules formed by the linking together of smaller molecules called nucleotides. A nucleic acid, deoxyribonucleic acid or DNA, is the material which comprises the gene.

Obtain the following pieces for your team. During the exercise, always keep the printed side of each piece up.

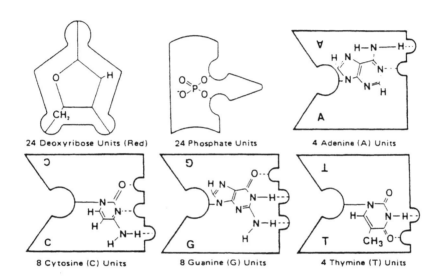

Figure 18.1

THE DNA NUCLEOTIDES

Examine a deoxyribose unit. Deoxyribose is a sugar. The three knobs represent covalent chemical bonds which deoxyribose can form with the bases and phosphate. The molecule consists of a ring of four carbon atoms and one oxygen atom. Individual carbon atoms are not shown, but are located at the intersections of lines. To one side of the oxygen atom is a carbon atom bonded to a CH_2 group. Attach a phosphate unit to the free bond that comes from the CH_2 group. On the opposite side of the oxygen atom is a carbon atom with a free bond shown. Attach an adenine base to this bond. The structure now represents an adenine nucleotide. The representation is incomplete: The adenine nucleotide is a triphosphate, not a monophosphate as shown by your representation. The two additional phosphate groups are removed during DNA synthesis. The phosphate end of the molecule is referred to as the 5' end. The as yet unused covalent bond comes from the 3' end of the molecule. Put together the rest of the deoxyribose, phosphate, and base units to form additional nucleotides. After all the nucleotides have been assembled, set aside one-half of *each* type nucleotide for later use in DNA replication.

Reprinted by permission of Carolina Biological Supply Co.

COVALENT BONDING OF NUCLEOTIDES

Select an adenine nucleotide and a guanine nucleotide. Join them by connecting the phosphate unit of one to the free carbon bond of the other. Continue adding nucleotides in this manner until you have built up a strand of 6 nucleotides as shown in Figure 18.2. The actual linkage of nucleotides is an enzyme controlled process which involves splitting off phosphate groups as noted above.

BASE PAIRING

Select one of the remaining nucleotides and attempt to pair it with a nucleotide of the strand by putting the ends of their bases together. Attempt several such pairings using different bases. Then fill in the following: Adenine pairs only with _____. Cytosine pairs only with _____.

Bases which will pair are said to be complementary. The bonds between the bases are not covalent bonds, but much weaker *hydrogen bonds*.

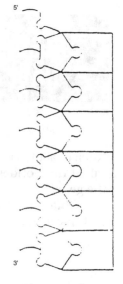

Figure 18.2

DNA STRUCTURE

Duplicate Figure 18.3 by pairing each nucleotide of the first strand with its complementary nucleotide and joining the bonds to form a new strand. Notice that the two strands are antiparallel (they run in opposite directions); that is, if one strand runs 5′ to 3′ (phosphate-sugar, phosphate-sugar) from top to bottom, the other must run 3′ to 5′ from top to bottom. The structure which is produced represents the double-stranded DNA molecule. Using A, C, G, T indicate on Figure 18.3 the base sequence which you have produced for each side of the molecule.

THE GENETIC CODE

The sequence of bases in DNA forms a code or set of instructions for protein synthesis. Each code word or *codon* consists of a sequence of three bases and specifies a particular amino acid. There are 64 possible codons (Fig. 18.4). Some are chain terminating codons which are not known to specify any amino acid. These indicate where a protein chain is to end. In most cases, an amino acid can be specified by more than one codon, but each codon can specify only one amino acid. Usually one strand (the sense strand) of DNA is active in protein synthesis.

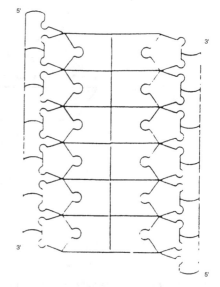

Figure 18.3

The codons of the inactive or anti-sense strand are not used in protein synthesis. Consider the right side of the molecule that you have constructed to be the sense strand and read the codons from the 3′ end of the strand toward the 5′ end. Which amino acids are specified by the two codons? _____ and _____.

DNA REPLICATION

Separate the two DNA strands to a distance of about 30 cm. The weak hydrogen bonding between the base pairs allows for this separation. From the free nucleotides left on your work space, rebuild the missing half of each DNA strand. Begin at the 3′ end of the original strand and proceed to its 5′ end. Compare the base sequences of the new molecules to the base sequence of the original molecule as recorded on Figure 18.3. Has the original base sequence been maintained during replication? _____.

104 Exploring Biology in the Lab

DNA CODON (3-5)	AMINO ACID SPECIFIED
ATT ATC ACT	CHAIN TERMINATING
CGA CGG CGT CGC	Alanine
GCA GCG GCT GCC TCT TCC	Arginine
TTA TTG	Asparagine
CTA CTG	Aspartic Acid
ACA ACG	Cysteine
GTT GTC	Glutamine
CTT CTC	Glutamic Acid
CCA CCG CCT CCC	Glycine
GTA GTG	Histidine
TAA TAG TAT	Isolecine

DNA CODON (3-5)	AMINO ACID SPECIFIED
AAT AAC GAA GAG GAT GAC	Leucine
TTT TTC	Lysine
TAC	Methionine
AAA AAG	Pherviaianine
GGA GGG GGT GGC	Protine
AGA AGG AGT AGC TCA TCG	Serine
TGA TGG TGT TGC	Threonine
ACC	Tryplophan
ATA ATG	Tyrosine
CAA CAG CAT CAC	Valine

Figure 18.4

REVIEW

Repeat the above procedures until you are familiar with DNA structure and replication.

Name _____ Section _____ Date _____

Exercise 18, Part 2 — Student Worksheet
RNA: The Code Transcribed

RNA is produced by using DNA as a pattern. During this process, called transcription, the genetic code is transferred from DNA to RNA.

Obtain the following units for your team; 12 deoxyribose (red), 12 ribose (pink), 24 phosphate, 4 adenine (A), 8 cytosine (C), 8 guanine (G), 2 thymine (T), 2 uracil (U). During the exercise, always keep the printed side of each piece up.

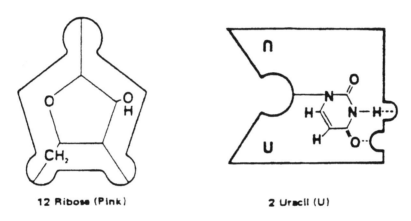

Figure 18.5

NUCLEOTIDES

Compare a deoxyribose unit to a ribose unit. Look at the formula of each molecule as printed on the units. How does deoxyribose differ from ribose sugar? _____
_____ .

The nucleotides of DNA contain deoxyribose sugar while nucleotides of RNA contain ribose sugar.

Use the deoxyribose units to form the following DNA nucleotides: 2 adenine, 6 cytosine, 2 guanine, and 2 thymine. Now use the ribose units to form the following RNA nucleotides: 2 adenine, 2 cytosine, 6 guanine, and 2 uracil.

Reprinted by permission of Carolina Biological Supply Co.

107

TRANSCRIPTION

During transcription the two strands of a DNA molecule become separated along part of the molecule's length. One strand of DNA remains inactive, but the other, the sense strand, is used to bring about the synthesis of RNA. To demonstrate transcription, first use the deoxyribose nucleotides to form a single strand of DNA. We will use this as the sense strand and ignore the antisense strand. Beginning at the 3' (sugar) end of the strand, pair each DNA nucleotide with its RNA complement (uracil of RNA complements adenine of DNA) and link together the phosphate and ribose units to form a strand of RNA. (Notice that RNA synthesis begins at the 5' end of the RNA strand and proceeds toward its 3' end.) Separate the RNA and DNA strands. Once RNA is formed, it is released from DNA and goes to other parts of the cell. How does transcription differ from replication? _____

_____.

_____.

TYPES OF RNA

Transcription produces three major types of RNA: ribosomal (rRNA), transfer (tRNA), and messenger (mRNA). Ribosomal RNA combines with proteins to form bodies called ribosomes. The ribosomes become located in the cell's cytoplasm and are centers for protein synthesis. Transfer RNA reacts with amino acids in the cytoplasm and interacts with ribosomes and messenger RNA in the assembly of amino acids into proteins. It is messenger RNA that brings the instructions for protein synthesis (the genetic code) from DNA in the nucleus to the ribosomes.

MESSENGER RNA AND THE GENETIC CODE

Look again at the sense strand of DNA and the RNA strand produced by it. Notice that the sequence of bases in RNA is controlled by the sequence of bases in DNA; that is, the RNA codons (triplets of bases) are complementary to the DNA codons. Figure 18.6 gives the possible DNA codons with their corresponding mRNA codons and the amino acids specified by each. (In the table, the mRNA codons are read from the 5' end of the molecule to the 3' end.) Which amino acids are specified by the RNA strand that you have assembled?

_____, _____, _____, and _____

DNA CODON (3-5)	mRNA CODON (5-3)	AMINO ACID SPECIFIED
ATT	USS	CHAIN TERMINATING
ATC	UAG	
ACT	UGA	
CGA	GCU	Alanine
CGG	GCC	
CGT	GCA	
CGC	GCG	
GCA	CGU	Arginine
GCG	CGC	
GCT	CGA	
GCC	CGG	
TCT	AGA	
TCC	AGG	
TTA	AAU	Asparagine
TTG	AAC	
CTA	GAU	Aspartic Acid
CTG	GAC	
ACA	UGU	Cysteine
ACG	UGC	
GTT	GAA	Glutamine
GTC	GAG	
CTT	GAA	Glutamic Acid
CTC	GAG	
CCA	GGU	Glycine
CCG	GGC	
CCT	GGA	
CCC	GGG	
GTA	CAU	Histidine
GTG	CAC	
TAA	AUU	Isolecine
TAG	AUC	
TAT	AUA	

DNA CODON (3-5)	mRNA CODON (5-3)	AMINO ACID SPECIFIED
AAT	UUA	Leucine
AAC	UUG	
GAA	CUU	
GAG	CUC	
GAT	CUA	
GAC	CUG	
TTT	TTT	Lysine
TTC	TTC	
TAC	TAC	Methionine
AAA	UUU	Phenyialanine
AAG	UUC	
GGA	CCU	Protine
GGG	CCC	
GGT	CCA	
GGC	CCG	
AGA	UCU	Serine
AGG	UCC	
AGT	UCA	
AGC	UCG	
TCA	AGU	
TCG	AGC	
TGA	ACU	Threonine
TGG	ACC	
TGT	ACA	
TGC	ACG	
ACC	UGG	Tryplophan
ATA	UAU	Tyrosine
ATG	UAC	
CAA	GUU	Valine
CAG	GUC	
CAT	GUA	
CAC	GUG	

Figure 18.6

REVIEW

Repeat the above procedure until you are familiar with the process of transcription.

Name _____ Section _____ Date _____

Exercise 18, Part 3 — Protein Synthesis: The Code Translated

Messenger RNA (mRNA) and transfer RNA (tRNA) work together to position amino acids at the ribosome for protein synthesis. The kinds and numbers of amino acids used are determined by the mRNA codons; thus, the genetic code is translated into a sequence of amino acids during protein synthesis.

Obtain the following units: 12 deoxyribose, 12 ribose, 24 phosphate, 2 alanine tRNA, 2 glycine tRNA, 2 alanine, 2 glycine, 2 alanine activating, 2 glycine activating, 4 adenine (A), 8 cytosine (C), 8 guanine (G), 2 thymine (T), and 2 uracil (U). Also obtain a ribosome template. During the exercise always keep the printed side of each piece up.

TRANSCRIPTION

Use the deoxyribose units to form the following DNA nucleotides; 2 adenine, 6 cytosine, 2 guanine, and 2 thymine. Now use the ribose units to form the following RNA nucleotides: 2 adenine, 2 cytosine, 6 guanine, and 2 uracil.

Use the deoxyribose nucleotides to form a single strand of DNA identical to that shown in Figure 18.7. The base sequence (beginning at the 3' end of the molecule) should be CGT, CCA, CGT, CCA as shown. Now transcribe a strand of mRNA from the DNA strand.

TRANSFER RNA

Each tRNA molecule (there is a different tRNA for each kind of amino acid) consists of about 75 nucleotides. One loop of the molecule has three unpaired bases that are complementary to one of the codons of mRNA. These three bases (called an anticodon) are shown magnified at the top of the tRNA units. One end of the tRNA molecule terminates in the nucleotide sequence CCA. An activating enzyme can attach an amino acid to the terminal A nucleotide of the tRNA. The enzyme is specific for a particular amino acid and for a particular tRNA. Thus, the amino acid glycine can only be attached to a tRNA specific for glycine.

Attach each of the activating units to the amino acid unit for which it is specific. Then attach each activating unit-amino acid complex to a tRNA unit.

PROTEIN SYNTHESIS

Position the mRNA on the mRNA binding site of the 40 S subunit of the ribosome template. The 5' (phosphate) end of the mRNA should be to the right as indicated on the sheet. Pick up a tRNA-amino acid unit whose anticodon complements the mRNA codon above the P site of the 60 S subunit and place it on the P site with codon and anticodon paired. Do the same for the A site.

Detach the P site amino acid from its tRNA activating unit and slide it over the tRNA, attaching it to the A site amino acid. Move the mRNA to the right until the tRNA-activating unit amino acid chain occupies the P site (the tRNA-activating unit amino acid chain is moved from the A site to the P site). Carefully lift the empty tRNA-activating unit complex, slide it out from under the amino acid chain, and place it to one side. Fill the A site with another tRNA-amino acid unit and repeat the above process until you have produced a chain of the four amino acid units. Then fill in Table 18.1.

Reprinted by permission of Carolina Biological Supply Co.

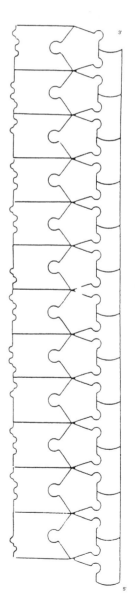

Figure 18.7

Table 18.1

DNA Codons (3' End)		mRNA Codons (5' End)		Amino Acids in Protein Chain
CGT	→	_____	→	_____
CCA	→	_____	→	_____
CGT	→	_____	→	_____
CCA	→	_____	→	_____

Suppose you had started with the DNA codons CGT, CCA, CCA, CGT. How would this have affected the outcome? _____

_____

_____

REVIEW. Repeat the above procedure until you are familiar with protein synthesis.

Name _____ Section _____ Date _____

Exercise 19 Macromolecular Model Lab

OBJECTIVES

After completing this exercise, you should be able to:

1. Construct and replicate a DNA molecule.
2. Use this DNA molecule to construct a simple protein.

MATERIALS

Model kits with directions (2 students/kit) Constructed model, if available

METHOD

1. Count pieces of kit
2. Identify the following pieces:
   - A = Deoxyribose Sugar
   - B = Ribose Sugar
   - C = Phosphate
   - D =
   - E =
   - F =
   - Bases G =
   - J =
3. DNA replication
   a. Construct a DNA molecule consisting of 6 base pairs
      (*Note:* Use a *variety* of nucleotides!)
   b. Construct a DNA nucleotide pool consisting of *at least 12* nucleotides
   c. Demonstrate DNA replication to your instructor, using your DNA nucleotide pool
4. RNA/Protein Synthesis
   a. Construct a DNA molecule consisting of 6 base pairs
   b. Construct an RNA nucleotide pool consisting of *at least 6* nucleotides
   c. Construct t-RNA structures (anticodon regions only) with attached amino acids
   d. Demonstrate to your instructor the construction of a mRNA molecule of at least 6 bases against *one side* of your DNA molecule
   e. Using this mRNA, demonstrate to your instructor the synthesis of a simple protein
   f. Answer the following questions:
      (1) Where in the cell life cycle does this occur?
      (2) What happens to the mRNA?
      (3) What happens to the tRNA?
      (4) What are the amino acids that make up your protein?

Name _____ Section _____ Date _____

Exercise 20 Genetics

Each egg and sperm normally contains a single set of chromosomes and is referred to as being **haploid.** Following fertilization, a set of chromosomes from both the sperm and the egg have been brought together in the zygote. The latter, containing two sets of chromosomes, is **diploid.** The zygote commonly divides by mitosis to produce two diploid daughter cells. These cells as well as all of the succeeding cells which continue to be produced by mitosis form a body structure composed of diploid cells. The haploid condition returns when certain specific cells undergo **meiosis** to produce **gametes:** eggs (in the female) or sperm (in the male).

Each chromosome of the haploid set has a characteristic appearance and contains a complement of genes which determine a set group of characters. One particular gene normally occupies a given position in the length of the chromosome. This position is called the **locus** of that gene. When the diploid condition is established, the nucleus now contains two chromosomes of each kind, i.e., one set from each of the two gametes. The one normal exception is in the case of the X and Y (sex) chromosomes. The chromosomes which can be thus paired—same appearance and gene complement—are referred to as **homologous chromosomes.** Genes that occupy the same locus on homologous chromosomes are called **alleles.** If both alleles are the same (have the same expression), the condition is called **homozygous;** if they are different, the condition is called **heterozygous.** The genes represented in the nucleus are collectively called the **genotype.** The detectable aspects (eye color, height, etc.) of the expression of the genes is called the **phenotype.**

During meiosis, homologous chromosomes pair (synapse) and then each member is distributed to a different haploid nucleus. The genes are distributed in the same manner as the chromosomes on which they are located. The result is gametes—each with a single set of chromosomes and one of the genes from each pair of alleles. The alleles are said to have **segregated.**

The behavior of genes discussed above may be summarized as in the following diagram, assuming the parents to be heterozygous (Aa).

Parents:	(Male)	Aa	x	Aa	(Female)
Meiosis:					
Types of Gametes:		A a		A a	

Fertilization:

		Female Gametes	
		A	**a**
Male	**A**	AA	Aa
Gametes	**a**	Aa	aa

Genotype ratio: 1 AA: 2 Aa: 1 aa

MONOHYBRID CROSS

Students, working in pairs, will attempt to reproduce the above ratio by a simple experiment. One student will toss the coins and the second student will record the results.

Take two different types of coins, such as a nickel and a penny. Have **A** equal a head and **a** equal a tail. Consider one of the coins (nickel) as the ovum and the other (penny) as the sperm. If a single coin is tossed a sufficient number of times, one would expect the appearance of an equal number of heads and tails. Or as related to the situation with the genotype in this experiment, there is an equal number of ova (or sperm) with the gene **A** as with the gene **a.** Flip both coins at the same time; and for each flip, record the combination: ***head–head, head–tail,*** or ***tail–tail.*** A total of 50 tosses should be adequate.

By assuming that the result of each coin toss represents the formation of a zygote, it is possible to determine the genotype of the zygote directly. Summarize your results to show the total number of each of the different combinations and the ratio of these combinations. In order to obtain larger and probably more significant numbers, the results of the class will be summarized on the blackboard. The ratio is calculated by dividing the smallest number into the other numbers. Then consider the smallest number as having a value of one in the ratio.

If one now assumes that gene **A** is dominant to gene **a,** i.e., the expression of **a** (recessive) is masked in the presence of **A,** the ratio of phenotypes becomes ***3:1.*** This is a basic phenotypic ratio in a situation where:

1. A single pair of alleles is involved.
2. Both parents are heterozygous.
3. One allele is dominant and the other is recessive.

Other characteristic ratios are obtained when more genes or different modes of gene expression are involved. Knowing this, it is possible to utilize this information and to work in the reverse direction. By observing the phenotypic ratio in controlled breeding experiments, one can determine how many alleles are involved and whether dominance plays a role.

The manner of breeding through the F_1 and F_2 generations to obtain a 3:1 ratio will be reviewed.

DIHYBRID CROSS: INDEPENDENT SEGREGATION

Again working in pairs, toss four coins (2 nickels and 2 pennies) simultaneously. Record your results for 50 tosses as well as the results of the class as follows:

Nickels	=	Pennies	Your Class Ratio
HH		HH	
HH		HT	
HH		TT	
HT		HH	
HT		HT	
HT		TT	
TT		HH	
TT		HT	
TT		TT	

Now assume that each nickel represents a single chromosome of one homologous pair and that each penny represents the same for a second homologous pair. Heads and tails can also be translated into genes.

Nickel: Head—A
Tail—a **A** is dominant to **a**
Penny: Head—B
Tail—b **B** is dominant to **b**

Having assumed the existence of dominance and recessiveness, regroup your data in this second set of coin tosses. This regrouping results in the establishment of a new phenotypic ratio. This is the typical phenotypic ratio produced in the F_2 generation when two pairs of alleles (showing dominance and recessiveness) are situated on different pairs of homologous chromosomes.

Would this same ratio be obtained if both pairs of alleles were located on the same pair of homologous chromosomes?

Exercise 21: Student Guide 17-3830 Human Chromosome Analysis BioKit®

You will first prepare a karyotype of a normal human male, and then you will have an opportunity to prepare an abnormal karyotype.

You will be given a blank Karyotype Form and a Biophoto® Sheet showing the chromosomes from a normal human male. Carefully cut out the individual chromosomes on the blank karyotype form. Do not fasten the chromosomes to the karyotype form until your instructor has checked it. Keep all paper scraps until you have identified each chromosome.

The chromosomes can be arranged in seven groups (A–G) according to length. Group A consists of the six longest chromosomes. The B group consists of four long chromosomes with the centromeres very close to one end. The C group consists of 14 medium-length chromosomes with the centromeres slightly off-center. The female sex chromosome (X chromosome) also falls into this group. Therefore a male will have 15 C-length chromosomes and a female will have 16. The D group consists of six chromosomes slightly smaller than the C's with the centromeres very near one end. The E group resembles the C group, but the chromosomes are much smaller. The F-group chromosomes are very small with the centromeres in the middle. The G group includes the four smallest chromosomes with the centromeres so close to the ends that it is difficult to see any short arms at all. The male sex chromosome (Y chromosome) falls into the G group. Therefore a male will have five G-length chromosomes and a female four.

It is possible to distinguish individual chromosomes. Using new staining techniques, each chromosome is stained to show the horizontal bands that are unique to that chromosome. Note the banding patterns of the chromosomes, usually about 12 bands per chromosome. Each band represents regions covering several hundred genes.

When you have completed the normal karyotype, you will be given a Biophoto® Sheet illustrating one of the following abnormal chromosome complements: Trisomy 21 (extra 21 chromosome) Down syndrome, 14/21 translocation carrier (a 21 chromosome attached to a 14 chromosome), 14/21 translocation Down syndrome, or 5p deletion (a loss of part of the short arms of the fifth chromosome, cat cry syndrome).

First, count the chromosomes on the Biophoto® Sheet; the normal number is 46. If there are more than 46 chromosomes, you know there is extra chromosomal material. If there are less than 46, there are two possibilities: either a chromosome has been lost, or two chromosomes have joined (translocation). Of the four abnormalities in this kit, the 14/21 translocation carrier has 45 chromosomes, the 5p deletion and the 14/21 translocation Down syndrome each has 46, and the trisomy 21 Down syndrome has 47. After counting the chromosomes, you can reasonably guess which abnormality your picture illustrates. Now cut out the chromosomes and prepare the karyotype. Remember, a female has two X chromosomes, while a male has one X and one Y chromosome. Each example in this kit has a normal number of sex chromosomes.

When you prepare the 14/21 translocation karyotype, a 14 chromosome and a 21 chromosome will be missing, but an extra chromosome similar to the C group will be present. This suggests that one of the chromosomes appearing to belong to the C group is actually a 21 chromosome attached to the short arm (p) of the 14 chromosome. Put the extra C-type chromosome with pair number 14 of the D group.

Copyright © 1985 by CAROLINA BIOLOGICAL SUPPLY CO. Reprinted by permission.

If your picture shows 46 chromosomes, it is either the 5p deletion or the 14/21 translocation Down syndrome. The karyotype for the 5p deletion will be normal, except one of the 5th chromosomes will have lost part of its short arms. The karyotype for the 14/21 translocation Down syndrome will have an extra C-type chromosome and will be missing a 14 chromosome. Put the extra C-type chromosome with pair number 14.

The trisomy 21 Down syndrome has 47 chromosomes; your karyotype should show an extra 21 chromosome. On the bottom of your completed karyotype list the number of chromosomes, the sex of your subject, and the abnormality (e.g., 47 chromosomes, female, trisomy 21 Down syndrome).

There are many kinds of chromosomal abnormalities other than the four used in this exercise. Not everyone with 45 chromosomes is a 14/21 translocation carrier, nor does everyone with 47 chromosomes have Down syndrome. Loss or gain of chromosomal material is frequently but not always associated with mental retardation. In the United States approximately 20,000 infants are born with chromosomal abnormalities each year; this is about 1 out of every 200 live births.

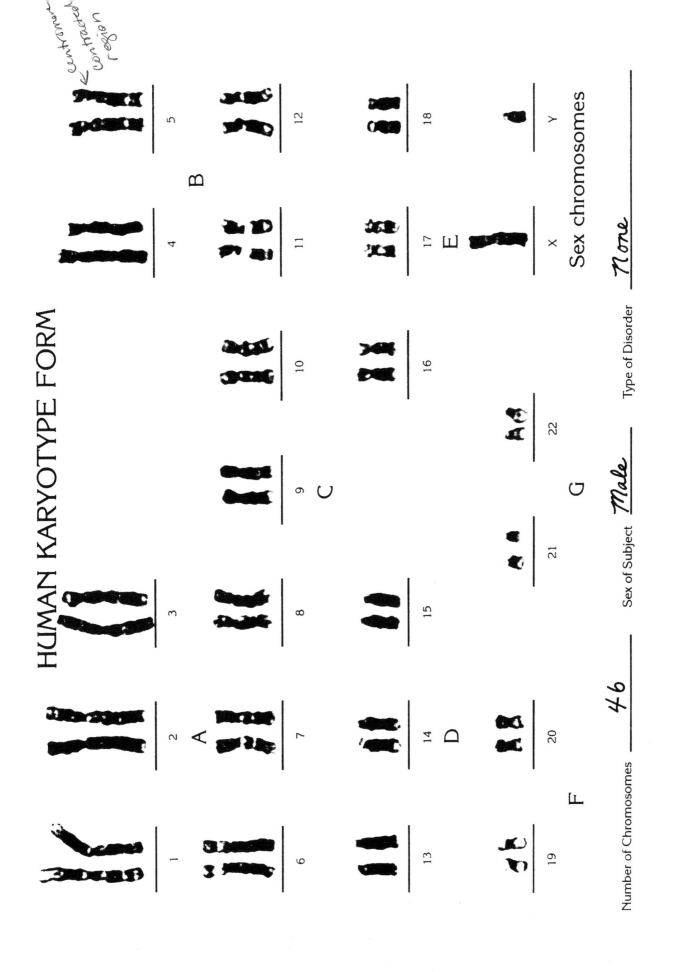

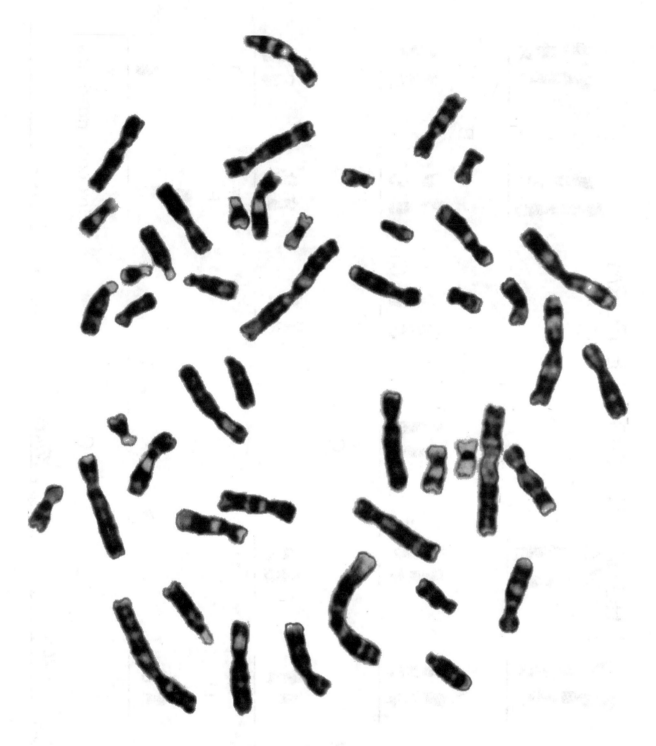

**Human Chromosome 1.
Normal Male 46,XY**

Name _____ Section _____ Date _____

Exercise 22: Determining the ABO-Rh Blood Type of Simulated Blood Samples

INTRODUCTION

Blood is a tissue comprised of four components: plasma, red and white blood cells, and platelets. Plasma is a clear straw-colored liquid portion that makes up 55% of the blood. It is comprised of a mixture of water, sugar, fat, protein, and various salts. In addition, plasma contains a number of blood clotting chemicals that help to stop bleeding.

Red and white blood cells and platelets make up the remaining 45% of the blood. Red blood cells or erythrocytes are tiny biconcave disks. Each red blood cell contains the oxygen binding protein hemoglobin. Hemoglobin contains four iron ions which bind with oxygen (O_2) and carbon dioxide (CO_2). The shape of red blood cells provide a greater surface area through which gases can diffuse and bind to the iron groups. The average red blood cell is about 7.5 μm in diameter and 2 μm in thickness (Fig. 22.1)

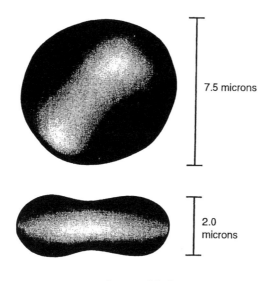

Figure 22.1
Red Blood Cell

Blood functions principally as a vehicle which transports gases, metabolic waste products, and hormones throughout the body. As blood passes through the lungs, oxygen molecules attach to the hemoglobin. As blood passes through the body's tissues in capillary beds, the hemoglobin releases the oxygen. Carbon dioxide and other waste gases are, in turn, transported by the hemoglobin back to the lungs. Thereafter, the process is repeated.

Surface proteins on red blood cells determine an individual's blood type. These surface proteins are called antigens. The system used to classify human blood is called the "ABO system." Dr. Karl Landsteiner, an Austrian physician, received the Nobel Prize in physiology for this discovery in 1930.

With the ABO system, the kind of antigens present on red blood cells determines the blood type. An individual with A antigens has blood type A, one with B antigens has blood type B, one with both A and B antigens has blood type AB, and one with no antigens on the surface of his/her red blood cells has type O.

From *Fundamentals of Biology Laboratory Manual* by Robert Keeton and Lewis Hunt. Copyright © 2004 by Kendall/Hunt Publishing Company. Reprinted by permission.

Blood plasma has circulating proteins called "antibodies." For example, individuals with A surface antigen have anti-B antibodies; those with B surface antigen have anti-A antibodies. Those with both A and B surface antigens have no antibodies. Individuals with no surface antigens have both anti-A and anti-B antibodies.

Blood typing is performed using "antiserum"—blood that contains specific antibodies. "Anti-A Serum," which contains anti-A antibodies, and "Anti-B Serum," which contains anti-B antibodies, are used in ABO blood typing. To perform a blood typing test, anti-A and anti-B sera are each separately mixed with a drop of sample blood and observed for "agglutination" or clumping.

Another important antigen on the surface of red blood cells is the Rh protein, named for the rhesus monkey in which it was first studied. People who have this protein are "Rh-positive," and those who lack it are "Rh-negative." Rh-negative individuals who have been transfused with Rh-positive blood can produce Rh antibodies. They may develop a transfusion reaction, during which agglutination may occur, if they are transfused again with Rh-positive blood; Rh compatibility is tested when the ABO blood type is determined.

Data Table 22.1
ABO Blood Types Summary

Blood Type	Antigen on Red Blood Cells	Antibodies in Plasma	Can Receive Blood from . . .	Can Donate Blood to . . .
A	A	B	O, A	A, AB
B	B	A	O, B	B, AB
AB	A and B	None	O, AB, A, B	AB
O	None	A and B	O	A, A, B, AB

Data Table 22.2
Agglutination Reactions in the ABO System

ABO Agglutination Reaction		Blood Type
Anti-A Serum	Anti-B Serum	
Agglutination	No Agglutination	A
No Agglutination	Agglutination	B
Agglutination	Agglutination	AB
No Agglutination	No Agglutination	O

Data Table 22.3
Rh Agglutination Reactions

Rh Agglutination Reaction	Rh Factor
Agglutination	+
No Agglutination	−

Data Table 22.4
Frequency of ABO Blood Types and Rh Factor in the U.S.

Blood Type	Frequency Percentage	Blood Type & Rh Factor	Frequency Percentage
A	42	A +	34
		A −	8
B	10	B +	8
		B −	2
AB	4	AB +	3
		AB −	1
O	44	O +	35
		O −	9

THE GENETICS OF BLOOD TYPES

The human blood types A, B, AB and O are inherited by multiple alleles. Multiple alleles refers to three or more genes that occupy a single locus on a chromosome. Gene I^A codes for the synthesis of antigen (agglutinogen A), gene I^B codes for the production of antigen B on the red blood cells and gene i (i°) does not produce any antigens. The phenotypes listed in the table below are produced by the combinations of the three different alleles, I^A, I^B, i°. When genes I^B and I^A are present in an individual, both are fully expressed. Both I^A and I^B are dominant over i°; the genotype of an individual with blood type O must be i°i°.

Phenotypes	Possible Genotypes
A	$I^A I^A$
	I^A (or I°)
B	$I^B I^B$
	I^Bi (I°)
AB	$I^A I^B$
O	ii (I°I°)

Use I^A for antigen A; I^B for antigen B; i or I° for no antigens present

Gene I^A is dominant over i (I°)

Gene I^B is dominant over i (I°)

AB blood type results when both genes $I^A I^B$ are present

The ABO blood groups and other inherited antigenic characteristics of red blood cells are often used in medico-legal situations involving identification or disputed paternity. In paternity cases a comparison of the blood groups of mother, child and alleged father may exclude the man as a possible parent of the child. For example, a child of blood type AB whose mother is Type A could not have as a father a man whose blood group is O. Blood typing does not prove that an individual is the father of a child; it merely indicates whether or not he is a possible parent.

ACTIVITY #1: BLOOD TYPING USING SIMULATED BLOOD

What you need—per student

	Anti-A Serum (simulated)
	Anti-B Serum (simulated)
	Anti-Rh Serum (simulated)
1	Blood typing tray
	Paper towels
4	Patient Blood Samples (simulated)
1 set	Stirring sticks (blue, green, and yellow)

Step 1—Have your students place 3 drops of Patient 1 Simulated Blood Sample in each well on their blood tray.

Step 2—Next, the students should place 3 drops of Anti-A Simulated Serum in Well A.

Step 3—The students should now place 3 drops of Anti-B Simulated Serum in Well B.

Step 4—Finally, have your students place 3 drops of Anti-Rh Simulated Serum in Well Rh.

Step 5—Have students use a separate stirring stick to mix the simulated blood and serum in each well for about 10 seconds.

Step 6—Have your students carefully examine each well to determine if the simulated blood in each well has clumped or agglutinated. Have them record their results and observations in Data Table 22.5

Step 7—The students should thoroughly rinse the tray and stirring sticks and repeat Steps 1–6 to type the remaining, simulated blood samples.

UNDERSTANDING YOUR RESULTS

Clumping indicates that simulated blood sample contains antigens that reacted against the antibodies in the typing serum that the students mixed it with.

Type A—If the blood in Well A is the only blood that agglutinates or clumps, then the blood sample you tested is type A blood.

Type B—If the blood in Well B is the only blood that agglutinates or clumps, then the blood sample you tested is type B blood.

Type AB—If the blood in both Well A and Well B agglutinates or clumps, then the blood sample you tested is type AB blood.

Type O—If the blood in both Well A and Well B does not agglutinate or clump, then the blood sample you tested is type O blood.

Rh—If the blood in Well Rh agglutinates or clumps, then the blood sample you tested is Rh-Positive blood.

Data Table 22.5

Simulated Blood Sample	Agglutination in Well A (+/–)	Agglutination in Well B (+/–)	Agglutination in Well Rh (+/–)	Blood Type
Patient 1				
Patient 2				
Patient 3				
Patient 4				

QUESTIONS

1. If your blood type is B, which antigens are present on your red blood cells? What if your blood type is A, type AB, or type O?

2. Based upon your results, which ABO blood type can Patient 1 receive safely? Patient 2? Patient 3? And Patient 4?

3. Which patient is considered a universal donor?

4. How could you determine if a blood sample is compatible to transfuse from one individual to another in an emergency situation, if blood typing serum is not available?

5. What would happen to a type O patient if he receives type A or B blood?

6. What are the consequences of Rh incompatibility?

7. Could a man with an AB blood type be the father of an O child? Explain.

8. Could a man with an O blood type be the father of an AB child? Explain.

9. Could a Type B child with a Type A mother have a Type A father? Explain.

10. What are the possible genetic combinations of an offspring when the blood types of the parents are A and B?

11. If Mr. Smith is Rh positive (homozygous) and he marries Ms. Brown, who is Rh negative, what are the chances for an Rh+ child? An Rh negative child?

12. Do the same problem as #11, but assume that Mr. Smith is Rh positive (heterozygous).

ACTIVITY #2: TAKING A CLOSER LOOK AT BLOOD

Physicians and other health care professionals regularly examine blood under the microscope to identify infections, blood cell abnormalities, and to count the various types of cells. The cells of the blood are of two classes: red blood cells (RBCs), or erythrocytes; white blood cells (WBCs), or leukocytes, which in turn are of many different types. Platelets, or thrombocytes, are also present as are cell fragments.

The red blood cells are tiny, round, biconcave disks, without nuclei, that average about 7.5 microns (0.003 in) in diameter. Red blood cells, as well as most white cells and platelets, are made by the bone marrow. The main function of the red blood cells is to transport oxygen from the lungs to the tissues. A healthy 70 kg (154 lb) man has about 5 L (1.3 qt) of blood in his body containing more than 25 trillion RBCs. The normal life span of RBCs in the circulation is only about 120 days. Worn out RBCs are removed by the spleen and liver where hemoglobin is recycled.

A number of conditions can be diagnosed based upon the red blood cell count. A high RBC level, a condition called "erythrocytosis," can be caused by smoking, living at high altitudes, or by disease. Low red blood cell levels, a condition called "anemia," can be due to a loss of blood, loss of iron, a vitamin deficiency, or other disease conditions.

Data Table 22.6
Normal and Abnormal Red Blood Cell Counts

	Red Blood Cells	
Count	**Male**	**Female**
Normal (at birth)	5.1 million cells per ul	4.5 million cells per ul
Normal (adult)	5.4 million cells per ul	4.8 million cells per ul
Anemia (low RBC count)	< 4.5 million cells per ul	< 4 million cells per ul
Erythrocytosis (high RBC count)	> 6.8 million cells per ul	> 6 million cells per ul

Leukocytes, or white blood cells, are considerably larger than red cells, have nuclei, and are much less numerous; only one or two exist for every 1000 red blood cells, and this number increases in the presence of infection. There are three types of leukocytes, all involved in defending the body against foreign organisms: granulocytes, monocytes, and lymphocytes. There are three types of granulocytes: neutrophilis (the most abundant), eosinophils, and basophils.

Platelets (thrombocytes) are tiny bits of cytoplasm, much smaller than the red blood cells, which also lack nuclei. They are normally about 30 to 40 times more numerous than the white blood cells. They are produced as fragments of the cytoplasm of the giant cells of the bone marrow—the megakaryocytes. The platelets' primary function is to stop bleeding. When tissue is damaged, the platelets aggregate in clumps as part of the clotting process.

Materials Needed Per Group

Normal human blood smear microscope slide
(Carolina Cat. No. RG-31-3158)

Human sickle cell anemia microscope slide
(Carolina Cat. No. RG-31-7374)

Acute lymphocytic leukemia smear microscope slide
(Carolina Cat. No. RG-31-7422)

Compound microscope

Step 1—Place the normal human blood smear microscope slide under a compound microscope. Examine the smear at 40X magnification and note the various cells types present. Depending on the stain used, erythrocytes will be pink; granules in basophils should be blue; in eosinophils bright red; and in neutrophils lilac. Lymphocytes should have a solid blue nucleus, while platelets are clusters that stain blue.

Step 2—Examine the shape of red blood cells, the different types of leukocytes, and platelets. View the details of cellular structure under an oil immersion lens.

Step 3—Compare the normal human blood smear microscope slide to the other blood smears with a blood disorder. Identify differences in the number and shape of the various blood cell types.

QUESTIONS

1. What does a low number of red blood cells indicate?

2. What does a high number of white blood cells indicate?

3. Describe how a normal red blood smear differs from the sickle cell anemia smear and acute lymphocytic leukemia smear microscope slide.

Name _____ Section _____ Date _____

Exercise 23: Human Genetic Traits

INTRODUCTION

Look about the room and you can see that each person is individually unique and different. To help us further realize that individuality is more than skin deep, we shall examine a few human characteristics known to be controlled by genes.

METHODOLOGY

Using Biophoto sheets illustrating typical common phenotypes in addition to others, tabulate the results of your investigation of fellow classmates of inherited human characteristics. Check your phenotype, give your genotype, and give the total number of students in the class who show each of the characteristics.

Genetic Traits Chart

Trait	Gene	Symbol	Trait
Free earlobe	E	e/e	Attached earlobe
Widow's peak	W	w/w	No widow's peak
Tongue roller	R	r/r	Cannot roll tongue
Straight thumb	H	h/h	Hitchhiker thumb
Bent little finger	B	b/b	Straight little finger
Long Palmar muscle	L	l/l	No long Palmar muscle
Pigmented iris	P	p/p	No pigmented iris
PTC taster	T	t/t	No PTC taster
Mid-digital hair	M	m/m	No mid-digital hair
Long index finger	I	i/i	Short index finger
Left thumb on top	TH	th/th	Right thumb on top
Long eyelash	S	s/s	Short eyelash
Dimples	D	d/d	No dimples
Short hallux	HA	ha/ha	Long hallux

Tabulation

Trait	Check Your Phenotype	Check Your Genotype	Check Students in Class
Free earlobe	_____	_____	_____
Attached earlobe	_____	_____	_____
Widow's peak	_____	_____	_____
No widow's peak	_____	_____	_____
Straight thumb	_____	_____	_____
Hitchhiker's thumb	_____	_____	_____

continues

Tabulation (continued)

Trait	Check Your Phenotype	Check Your Genotype	Check Students in Class
Bent little finger	_____	_____	_____
Straight little finger	_____	_____	_____
Long Palmar muscle	_____	_____	_____
No long Palmar muscle	_____	_____	_____
Pigmented iris	_____	_____	_____
No pigmented iris	_____	_____	_____
Dimples	_____	_____	_____
No dimples	_____	_____	_____
Short hallux	_____	_____	_____
Long hallux	_____	_____	_____
Tongue roller	_____	_____	_____
Cannot roll tongue	_____	_____	_____
PTC taster	_____	_____	_____
No PTC taster	_____	_____	_____
Mid-digital hair	_____	_____	_____
No mid-digital hair	_____	_____	_____
Long index finger	_____	_____	_____
Short index finger	_____	_____	_____
Left thumb on top	_____	_____	_____
Right thumb on top	_____	_____	_____
Long eyelashes	_____	_____	_____
Short eyelashes	_____	_____	_____
Blood Group			
Type A	_____	_____	_____
Type B	_____	_____	_____
Type AB	_____	_____	_____
Type O	_____	_____	_____
Rh (D)—positive	_____	_____	_____
Negative	_____	_____	_____

Genetic Variability of Human Traits

Tabulate the number of persons in the class whose hands remain up after each characteristic is called. Do this for *five* persons.

Characteristic Trait	First	Second	Number of Persons Third	Fourth	Fifth
1	____	____	____	____	____
2	____	____	____	____	____
3	____	____	____	____	____
4	____	____	____	____	____
5	____	____	____	____	____
6	____	____	____	____	____
7	____	____	____	____	____
8	____	____	____	____	____
9	____	____	____	____	____
10	____	____	____	____	____
11	____	____	____	____	____
12	____	____	____	____	____

1. What is the average number of characteristics which must be considered before all hands go down?

2. Suppose you made a study of an equal number of people at a family reunion. Would you expect it to require more or less characteristics before all hands went down? Explain.

*(Optional)

*3. Below write out the expansion of the binomial $(P + q)^2$ as given by your instructor and make the proper substitutions to get the estimated percentages of homozygous dominant, recessive, and heterozygous persons in the class. Show all your calculations below. A possible characteristic to start with is to compare the percentages between the tasters and non-tasters.

Exercise 24: Genetic Problems

SET 1

1. In summer squashes white fruit color is dominant over yellow. If a squash plant homozygous for white is crossed with one which is homozygous for yellow, what would be the appearance of the F_1 generation? Of the F_2 generation? Of the offspring of a cross of the F_1 back on its white parent? Of the offspring of a cross of the F_1 back on its yellow parent?

2. In man, right-handedness (R) is dominant over left-handedness (r). A right-handed man whose mother was left-handed marries a right-handed woman whose father and two of his three sisters are left-handed. What chance will the children of this marriage have of being left-handed?

3. In summer squash, white fruit (W) is dominant over yellow (w), and disc shape (D) is dominant over sphere shape (d). In a cross between a squash plant homozygous for yellow fruit color and disc fruit shape, and one homozygous for white fruit color and sphere fruit shape, what will be the appearance, as to color and shape of fruit, of the F_1 generation? Of the F_2 generation? Of the offspring of a cross of the F_1 back on the yellow disc parent? On the white sphere parent?

4. What are the gametes formed by the following squash plants, and what will be the appearance of the offspring from each cross?

    WWdd × wwDD   WwDd × Wwdd
    WwDd × wwdd   WwDd × wwdd
    WwDd × WwDD   WwDd × WwDd

5. A brown-eyed, right-handed man marries a blue-eyed, right-handed woman. Their first child is blue-eyed and left-handed. If other children are born of this couple, what will probably be their appearance as to these two traits?

6. How would you recognize a line of garden peas that had become genotypically pure for a given trait?

7. A cross of two pink flowered plants produces offspring whose flowers are red, pink, or white. Defining your genetic symbols, give all the different kinds of genotypes involved, and the phenotypes they represent.

8. What conclusions could you reach about the parents if the offspring had phenotypes in the following proportions?

    a. 3:1   b. 1:1   c. 9:3:3:1   d. 1:1:1:1

9. In snapdragons, red flowers (R) are incompletely dominant to white (r), the hybrid being pink; narrow leaves (N) are incompletely dominant to broad leaves (n), the hybrid being intermediate in width ("medium"). Show the genotypes and phenotypes for the progeny of a cross between the following:

    a. red medium and pink medium plant
    b. a pink medium and white narrow
    c. two identical dihybrids

Exercise 24: Genetic Problems

SET 2

1. In guinea pigs, short is dominant to long. A short-haired guinea pig was mated to a long-haired one. What proportions of the offspring (F_1) will be expected to be:
   a. homozygous short-haired
   b. homozygous long-haired
   c. heterozygous short-haired
   d. heterozygous long-haired

2. A husband and wife both have normal skin pigmentation. Their first child is an albino. Give the genotypes of the parents and of the albino child. What is the chance that if they have a second child, this child will be an albino? What is the chance that if they have a third child, this child will be an albino?

3. Translate the following sequence of messenger RNA nucleotides in a sequence of amino acids forming part of a polypeptide chain:
   a. C-U-G-U-U-U-U-G-C-A-G-U-G-G-U-U
   b. Write the sequence of nucleotides for the portion of a DNA molecule that served as a pattern or template for the formation of the messenger RNA in the above.
   c. A certain protein contains the following sequence of amino acids: isoleucine-serine-arginine-glutamic acid-serine-proline-valine-glutamic acid. Using the table found in your text, write the sequence of RNA triplet that would code for this sequence of amino acids. Then write the sequence of DNA triplets that would code for the formation of RNA.

4. In man, right-handedness (R) is dominant over left-handedness (r). A right-handed man whose mother was left-handed marries a right-handed woman whose father and two of his three sisters are left-handed. What chances will the children of this marriage have of being left-handed?

5. In summer squash, white fruit (W) is dominant over yellow (w), and disc shape (D) is dominant over sphere shape (d). In a cross between a squash plant homozygous for yellow fruit color and disc fruit shape, and one homozygous for white fruit color and sphere shape, what will be the appearance as to the color and shape of the fruit. Of the F_1 generation? Of the F_2? Of the offspring of a cross of the F_1 back on the yellow disc parent? On the white sphere parent?

6. What are the gametes formed by the following squash plants, and what will be the appearance of the offspring from each cross?
   a. WWdd × wwDD
   b. WwDd × Wwdd
   c. WwDD × wwdd
   d. WwDd × wwdd
   e. WwDd × WwDD
   f. WwDd × WWDd

7. Two normal appearing parents produce a child suffering from sickle-cell anemia. What is the chance that the next child will also have this anemia?

8. Suppose that a group of people decide to migrate to an uninhabited island. A serologist tests them and finds that they are assorted:

    50 are type M

    300 are type MN

    150 are type N

    Are the blood types representative of a population in equilibrium? (Hardy-Weinberg)

9. Suppose two unrelated albinos married and had 8 children, 4 albino and 4 non-albino. How could you explain these results?

10. Outline briefly the "Operon Hypothesis."

11. A normal man marries a normal woman known to carry the hemophilia gene (a carrier). Hemophilia will likely occur in what percentage of their male children?

    a. Normal man × woman with hemophiliac father?

    b. Hemophiliac man × homozygous normal woman?

    c. Hemophiliac man × normal woman with hemophiliac father?

12. What will be the probable blood type of a child if:

    a. Parents belong to groups **AB** and **O**?

    b. Parents belong to groups **A** and **B**?

    c. Parents belong to groups **AB** and **B**?

    d. Parents belong to groups **O** and **O**?

13. Before mitosis begins, a human skin cell contains 46 chromosomes. When that cell is at a metaphase stage of mitosis, how many chromatids does it contain? How many centromeres? At the conclusion of mitosis, how many chromosomes does each of the two daughter cells contain? How many centromeres?

14. Assume organism X has 20 chromosomes and is undergoing spermatogenesis:

    a. How many chromatids are found in metaphase of the primary spermatocyte?

    b. How many centromeres?

    c. How many tetrads?

    d. How many chromatids does the secondary spermatocyte contain?

    e. How many centromeres?

    f. How many chromatids are found in the spermatids?

    g. Sperms?

15. Define, locate, or identify the following:

Purine	Allele	Codon
Locus	P_1 Generation	Operon
Dominant	Chiasmata	Recessive
Trisomy	Genotype	Phenotype
Feedback Inhibition	Dihybrid Cross	Jacob and Monod
5'–3', 3'–5';	Backcross	Sex Linked
One-gene-one-enzyme concept		Inborn error of metabolism

16. Diagram all phases of mitosis in plant and animals noting all of the structural similarities and/or differences.

17. Determine the gametes produced by each of the following genotypes:

    a. AaBb

    b. AaBbCc

    c. AaBbCcDd

    d. AaBbCcDdEe

    e. aa bb cc dd ee ff

18. Identify anomalies for each of the following: (Library Work)
    a. Klinefelter's syndrome
    b. Phenylketonuria
    c. Hypogammaglobulinemia
    d. Down's syndrome
    e. Tay Sach's syndrome
    f. Turner's syndrome
    g. Galactosemia
    h. Huntington's Chorea
    i. Ocular Albinism
    j. Alkaptonuria

19. In man, a type of blindness called aniridia (A) and migraine headache (M) are both dominant to the normal traits. What are the chances two AaMm people, both suffering from aniridia and migraine, would produce a normal child with neither parent trait?

20. Explain the following phenotypic ratios:
    a. 3:1
    b. 9:3:3:1

Exercise 24: Genetic Problems

SET 3

1. In ghostly goblins, ghastly glowing eyes are dominant over faintly flashing eyes. A glowing eyed male goblin, Godfrey, whose mother had flashing eyes marries Greta, who has glowing eyes and whose father and two sisters have flashing eyes. What are the chances that Godfrey and Greta will produce goblinets with flashing eyes?

2. In witches, stringy hair is dominant over smooth hair and missing teeth are dominant over a complete set of teeth. In a cross between a witch homozygous for string hair and complete teeth with a warlock homozygous for smooth hair and missing teeth what will the F_1 generation look like? The F_2 generation? An F_1 warlock mated with a witch homozygous for stringy hair and missing teeth? An F_1 witch mated with a warlock homozygous for smooth hair and complete teeth?

3. A glowing eyed green goblin marries a flashing eyed green goblin. Their first goblinet is golden and has flashing eyes. If the proud goblins have more goblinets, what will they probably look like?

4. Crossing two vampires with medium fangs produces offspring with long, medium, and short fangs. Defining your symbols, give all the different kinds of genotypes involved and the phenotypes they represent. What would happen if a medium fanged vampire mated with a short fanged vampire?

5. Lycanthropy is a disease that causes one to become a werewolf at the time of the full moon. It is caused by a recessive allele on the X chromosome. A boy has the disease; neither his parents nor his grandparents have the disease. What are the genotypes of the parents, grandparents, and the boy? Please draw a pedigree to indicate how the disease was inherited by the boy.

Part VII

Evolution

Name _____ Section _____ Date _____

Exercise 25: A Simulation of Natural Selection

Natural selection is recognized by nearly all biologists as the single most important mechanism leading to evolutionary change in a species. In spite of its widely recognized importance, the implications of the process of natural selection are poorly understood or misunderstood by many biologists and most nonbiologists. As you remember from lecture, natural selection is part of the Darwin-Wallace theory of evolution.

Whether or not a particular trait enhances an individual's fitness (ability to produce offspring) is entirely dependent upon the environment in which the individual lives. While red coloration may be favorable in one environment, it may be detrimental to reproduction in a different environment. Therefore, natural selection acting in different environments can cause originally identical populations to diverge and end up looking quite different.

The purpose of this exercise is to demonstrate the process of natural selection with some simple simulations, and to investigate the results of natural selection acting in different environments. We will use populations consisting of a variety of beans which will be scattered over selected backgrounds (environments). Students will act as predators on these bean populations. Seeds that escape predation will reproduce and the procedure will be repeated for 3 or 4 generations. The reproductive process will be asexual to avoid the necessity of simulating the genetic complexities of sexual reproduction. Thus, each survivor will produce offspring colored like itself.

PROCEDURE

1. Creating the population

    Students will work in groups of at least 3. Initial populations of 80 seeds should be created using 10 seeds of each of the 8 varieties. These varieties vary in color and shape, but are all approximately the same size. Each group will be assigned a different environment in which to perform the selection simulation. Each group will test one heterogeneous environment and one homogeneous environment.

2. Simulated natural selection

    One student from each group should act as the predator on the population. This student should gather beans one at a time (but as rapidly as possible), and drop them into the empty cup. This will ensure that the visual contact with the foraging area is broken momentarily following each capture. **Foraging must continue until 64 beans are captured, no more, no less.** Each foraging bout should be timed by another student of the team. This will allow an objective assessment of foraging efficiency from generation to generation. The same predator should do all the foraging in successive generations of the same environment.

    The exact number of beans taken must be carefully determined in order to know how many beans must be replaced to maintain a population of stable size. A fluctuating population introduces a variable we wish to avoid in the present study. All members of the team should count beans as they are removed by the predator. When exactly 64 beans have been taken, the predator should stop foraging and the number of each variety of beans taken should be determined and entered in the data table. To facilitate counting, place gathered beans in separate piles according to their variety. In this way it should be possible to quickly determine varieties that escaped detection and that sill remain in the habitat. The beans should then be placed back into the stock of unused beans. Please be sure to keep the bean varieties separate.

3. Reproduction

The number of survivors of each bean type can be determined by subtraction on the data table provided. **Each survivor will produce 4 identical offspring.** Simply count out four beans of the appropriate type of each bean left in the environment. Replication should be done with care, but if mistakes are made they will be regarded as mutations. The total number of offspring (64) should then be mixed and scattered over the environment, as was done with the initial generation.

Continue this procedure for the next 4 generations. After four rounds of predation it is not necessary to physically count out the offspring of the survivors. The number of offspring can simply be calculated by multiplying the number of survivors of each type of four. Remove the last 16 survivors before moving on to the next environment.

QUESTIONS TO CONSIDER

Think about the following questions. If you have time you may want to investigate them experimentally and test your predictions.

1. Are the results reproducible, either with the same predator or different predators?
2. What effect would a lower intensity of selection have on your results?
3. Would you expect the results of this simulation to vary with the structural complexity of the environment? What would be the results in a three-dimensional environment like leaf litter compared to the more or less two-dimensional environments used here?
4. How does foraging efficiency (time required to collect the 64 seeds) differ among generations and among environments? How would structural complexity affect foraging efficiency?
5. What would be the effect of changing the environments each generation, either gradually or by alternating contrasting backgrounds?
6. What would happen if some of the beans had slightly different reproductive values, for instance, if some consistently produced 5 offspring and the others produced 3?

Assignment

This is to be written following the scientific method. See the lab syllabus for specific directions. It is due typewritten next week.

Data Table for Natural Selection

Bean types

NB = navy beans
WK = white kidney beans
GB = garbanzo beans
HR = Habicheulas Rositas (pink beans)

RB = roman beans
RK = red kidney beans
CB = coffee beans
BB = black beans

Environment _____

	Bean Types								
Generations (G)	NB	WK	GB	HR	RB	RK	CB	BB	Total
G1—starting #	10	10	10	10	10	10	10	10	80
# captured									64 (time: _____)
# not captured									16
# of young = # not captured × 4									64
G2—starting # = # not captured + # of young									80
# captured									64 (time: _____)
# not captured									16
# of young									64
G3—starting #									80
# captured									64 (time: _____)
# not captured									16
# of young									64
G4—starting #									80
# captured									64 (time: _____)
# not captured									16
# of young									64
G5—starting #									80

Data Table for Natural Selection

Bean types
- NB = navy beans
- WK = white kidney beans
- GB = garbanzo beans
- HR = Habicheulas Rositas (pink beans)
- RB = roman beans
- RK = red kidney beans
- CB = coffee beans
- BB = black beans

Environment _____

	Bean Types								
Generations (G)	NB	WK	GB	HR	RB	RK	CB	BB	Total
G1—starting # # captured	10	10	10	10	10	10	10	10	80 64 (time: ____)
# not captured									16
# of young = # not captured × 4									64
G2—starting # = # not captured + # of young									80
# captured									64 (time: ____)
# not captured									16
# of young									64
G3—starting #									80
# captured									64 (time: ____)
# not captured									16
# of young									64
G4—starting #									80
# captured									64 (time: ____)
# not captured									16
# of young									64
G5—starting #									80

Part VIII

Survey of Kingdoms

Name _____ Section _____ Date _____

Exercise 26: Observation of Prokaryotic Cells

The Domain Bacteria includes two groups of prokaryotic organisms that we will look at in lab: the bacteria and the cyanobacteria.

STRUCTURAL FEATURES

Prokaryotes are single-celled organisms that lack a nucleus or other membrane-bound organelles. Their metabolic reactions take place at the plasma membrane. Their proteins are assembled on **70S ribosomes** in the cytoplasm. The majority of bacteria have a cell wall, usually containing a meshwork of cross-linked **peptidoglycan,** a tough carbohydrate.

Morphology

Bacterial cells come in one of three common shapes:

A. **Coccus**—spherical

B. **Bacillus** (rod)—cylindrical

C. **Spiral**—helical shaped; includes spirochetes and spirilla

Traditionally, bacteria have been partially identified by Gram staining: This procedure breaks bacteria into two major groups based on differences in their cell wall structure. The cell walls of **Gram-positive** bacteria retain the original dye, resist decolorization with alcohol, and appear a deep purple color under the microscope. **Gram-negative** bacteria lose the purple color of the original stain when washed with alcohol and stain pink with a counterstain. The Gram stain is the most commonly performed differential stain in microbiology. These two groups of bacteria have very different disease-causing properties and usually respond to very different antibiotics.

Exterior to the cell wall is the **glycocalyx,** a sugary capsule that helps pathogenic bacteria escape detection by the immune system and resist phagocytosis.

Two kinds of filamentous structures may be attached to the cell wall: The **bacterial flagellum** rotates like a propeller to propel the bacteria with a jerky motility. **Fimbria** and **pili** are projections that help bacteria attach to one another in conjugation, or help them attach to surfaces.

Metabolic Diversity

Bacteria are very metabolically diverse.

Photosynthetic autotrophs synthesize their own organic compounds using sunlight as the energy source.

Chemosynthetic autotrophs produce organic compounds using the energy in simple inorganic substances.

Heterotrophs must obtain their energy from organic compounds made by the photosynthetic and chemosynthetic autotrophs.

Bacterial Reproduction

Bacteria reproduce by *binary fission,* resulting in two genetically identical daughter cells. The mechanism is simpler than mitosis and meiosis because of the single chromosome—a circular DNA molecule.

Bacteria also contain small circles of DNA in their cytoplasm that are distinct from the main chromosome. These **plasmids** carry few (5–100) genes, but they may be genes for important characteristics such as antibiotic resistance or toxin production. Plasmids can also be exchanged between different bacteria—a process that contributes to the rapid spread of antibiotic resistance among bacteria.

Bacteria are widely distributed in the environment and can live in almost any environmental conditions, including the most extreme environments on the Earth. Some bacteria are able to persist in the environment because they form a dormant structure containing their DNA, called a spore. Spores from the wrappings of Egyptian mummies have been brought back to a feeding state after being dormant for thousands of years. Bacteria are important in the environment as decomposers of organic and inorganic compounds. They recycle nutrients such as carbon, nitrogen, sulfur, and potassium. Photosynthetic bacteria are important contributors to the Earth's atmospheric oxygen. Through the process of fermentation, bacteria are producers of many different types of foods such as cheeses, soy sauce, and sauerkraut. They are also the source of many antibiotics, which are antibacterial drugs. The normal bacteria that inhabit your body contribute to your overall state of health by inhibiting the growth of harmful bacteria, and providing digestive enzymes and Vitamin K. Although some bacteria cause important human diseases such as plague, tuberculosis, and Lyme disease, only about 10% of bacteria are pathogenic.

Part A—Observation of Bacterial Slides

Using the prepared, stained slides provided by your instructor, find and draw the three bacterial morphologies: coccus., bacillus, and spiral (Figure 26.1). Your instructor may provide you with additional slides to observe other structures such as flagella or bacterial spores.

Part B—Observation of Cyanobacteria

The cyanobacteria are a group of photosynthetic bacteria that were formerly called the blue-green algae. These are true prokaryotes, and although they are photosynthetic, they do not contain chloroplasts. Photosynthesis is carried out on the plasma membrane. They do contain photosynthetic pigments (chlorophyll) which gives them their characteristic color. In this exercise, we will look at two cyanobacteria: *Anabaena* and *Oscillatoria*.

Prepare a wet mount of each organism using the live cultures provided. Your instructor may also direct you to use prepared, fixed slides. Draw and label each organism (Figure 26.2). Use your atlas for additional reference material.

Part C—Culturing Microbes from the Environment

The purpose of this exercise is to demonstrate that bacteria are present throughout the environment. For this lab, you will sample some environmental surfaces as well as a body surface. You will grow the bacteria in Petri dishes that contain a semi-solid material called agar. Agar is a carbohydrate obtained from red algae; it makes an ideal solid growth medium for bacteria, allowing us to see bacterial colonies.

1. Obtain the following materials from your instructor:

   Three agar plates

   Two sterile cotton swabs

   A tube of sterile water

2. Be sure to label all plates with your lab group, and the name of the site that you sampled. Label the part of the dish that contains the agar—not the lid.

3. Expose one agar plate to the environment by leaving the lid off the plate and placing it with the agar surface facing up. Choose an environment that you suspect will contain bacteria, such as on top of the refrigerator, or beneath the lab bench. Leave your plate exposed for about 45 minutes. Then place the lid back on the plate and place it in the basket to be incubated.

4. Using a sterile cotton tip swab, sample an environmental surface by rubbing the swab over the area. If your choice of surface is dry, you may moisten the swab in the water before taking the sample. After taking the sample, rub the swab lightly across the agar surface of the Petri dish. Place the lid back on the plate and place it in the basket to be incubated.

5. Using another cotton swab, sample a body surface that you suspect might have bacteria present. The back of the throat, the gum line, the skin behind the ears, and the inside of the cheek are good sites to sample. DO NOT stick the swab in your ear, up your nose or in your eye. After taking the sample, transfer the material by rubbing the swab on the agar surface of the Petri dish. Place the lid back on the plate and place it in the basket to be incubated. DISCARD THE SWAB IN THE BIOHAZARD WASTE CONTAINER.

6. NEXT LAB PERIOD: Observe the plates for the growth of bacteria. Be careful handling the bacterial plates. You should wear gloves if you have any cuts or broken skin. Discard all plates in the biohazard bags when you are finished.

7. For each plate record the following: Area sampled, Number of bacterial colonies, Number of *different* bacterial colonies:

Area Sampled	Number of Colonies	Number of Different Colonies

Questions:

Which area contained the most bacteria. Can you explain why?

Which area contained the greatest diversity of bacteria. Can you explain why?

Which area contained the fewest number of bacteria. Can you explain why?

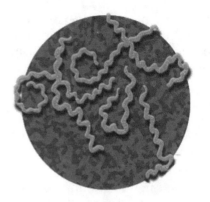  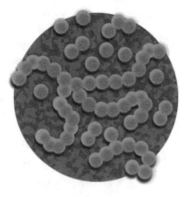

Spirillum (corkscrew-shaped)

Bacillus (rod-shaped)

Coccus (spherical)

Figure 26.1
Shapes of Bacteria

© Kendall/Hunt Publishing Company

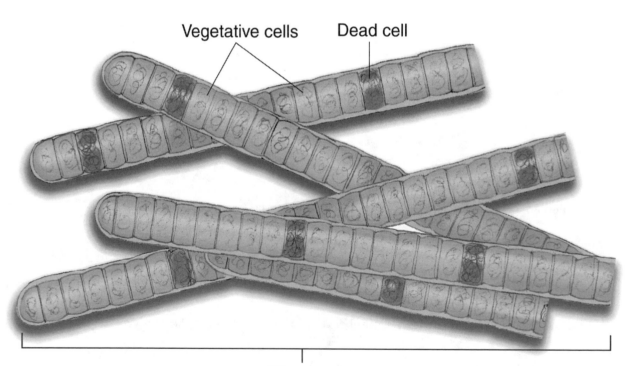

Figure 26.2
Cyanobacteria

© Kendall/Hunt Publishing Company

Name _____ Section _____ Date _____

Exercise 27 A Simulated Epidemic

INTRODUCTION

In the laboratory today, you will start a simulated epidemic and trace the pattern of its spread. You will use a fluorescent material that will represent a possible pathogenic organism. Identification of the fluorescent material will indicate a contaminated surface or individual. This is a laboratory exercise that will involve individual as well as class participation.

MATERIALS

- 1 container with 2 gumdrops/student
- Battery operated UV lamps
- Box of Kim-wipes
- Plastic bag for Kim-wipe disposal
- 4 small nail/hand scrub brushes
- Adequate supply of soap and paper towels

Methods

1. Students will thoroughly wash and dry hands, then examine hands under UV light and record results.
2. When instructed, students will pick up gumdrops in RIGHT HAND and gently roll candy around in hand. Return candy to foil dish. Using a Kim-wipe, GENTLY wipe hand that held the gumdrops. Dispose of Kim-wipe as instructed.
3. Examine hands with UV light and record results.
4. Students will shake hands with each other, using a pattern determined by the instructor. Pattern will be recorded and each student will receive a copy for the lab report.
5. After shaking hands, students will again examine hands under the UV light and record results.
6. Students will then be instructed to thoroughly wash and scrub their hands.
7. Students' hands, workstations, and surrounding areas will be reexamined with UV light. Observations will be recorded.
8. All materials will be disposed of as indicated by the instructor.

Note: Results should be recorded as follows:

"–"—no evidence of fluorescent "organisms"

"+"—evidence of fluorescent "organisms"

Indicate quantity of "organisms" using the following scale:

+ — slight/few "organisms" to ~25% coverage of surface

++ — ~50% of surface covered with "organisms"

+++ — ~75% of surface covered with "organisms"

++++ — entire surface covered with "organisms"

Discussion Points

1. Identify, with documentation, who started the "epidemic."
2. Describe, with documentation, how the "organism" was spread throughout the population.
3. Name 1 microscopic organism that might be spread through a population in this way and describe what effect this organism might have on infected members of the population.
4. Define "epidemic" and describe how epidemics occur.
5. Referring to our simulated epidemic, how might such epidemics be interrupted and controlled!

Simulated Epidemic
Results Record

Name	After Initial Hand Wash	After Handling Gum Drops	After Hand Shaking with Classmates

Name _____ Section _____ Date _____

Exercise 28, Part 1: Kingdom Protista—The Algae

PART I

OBJECTIVES

After completing this exercise, you should be able to:

1. Outline the major characteristics of different groups of algae.
2. Understand the relationship between the algae and members of the plant kingdom.
3. Describe the processes involved in the life cycle of *Chlamydomonas*.

INTRODUCTION

The algae consist of several groups of diverse organisms that use photosynthesis as their mode of nutrition. Algae have **eukaryotic** cells with nuclei, organelles, and cell walls. Their forms range from single cells **(unicellular)**, to chains of cells **(filamentous)**, to groups of cells attached to each other in a nonfilamentous manner **(colonial)**, to a more complex multicellular design, with plant-like bodies called **thalli**. The habitats of algae are varied, and most are found in a fresh or saltwater environment. Some can live on tree bark, rocks, in soil, or may be found in mutualistic relationships with other organisms.

The major phyla of algae can be characterized by their color, cell arrangement, photosynthetic pigments, cell wall, and storage products. These features are summarized in Table 28.1. The availability of light and the presence of certain photosynthetic pigments can determine where algae are located.

Table 28.1

	Color	Cell Arrangement	Pigments	Cell Wall	Storage Products
Dinoflagellates	Brown	Unicellular	Chlorophyll a and c, carotene, xanthins	Cellulose, silica	Starch
Euglenoids	Green	Unicellular	Chlorophyll a and b, carotene	None; protein pellicle	Glucose polymer
Diatoms	Brown	Unicellular	Chlorophyll a and c, carotene, xanthophylls	Pectin and silica	Oil
Brown Algae	Brown	Multicellular	Chlorophyll a and c, fucoxanthin	Cellulose, alginic acid	Carbohydrate (laminarin)
Red Algae	Red	Most are multicellular	Chlorophyll a and d, phycobilins	Cellulose	Starchlike
Green Algae	Green	Unicellular, multicellular	Chlorophyll a and b	Cellulose	Starch

From *Biological Investigations* by Gayne Bablanian. Copyright © 2004 by Kendall/Hunt Publishing Company. Reprinted by permission.

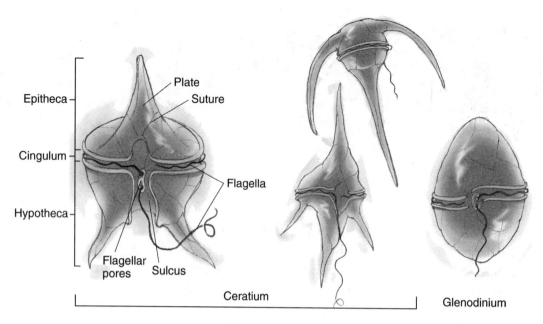

Figure 28.1
Dinoflagellates

© Kendall/Hunt Publishing Company

PHYLUM DINOFLAGELLATA

Dinoflagellates are unicellular **planktonic,** or free-floating, algae (Figure 28.1). Their cell walls are composed of many individual plates made of cellulose and silica. Two flagella attached to perpendicular grooves enable this phylum to be motile. Dinoflagellates are important primary producers in oceans. Some species are bioluminescent, while others **(zooxanthellae)** live in a symbiotic relationship with corals. Dinoflagellates of the genus *Gonyaulax* produce neurotoxins that cause paralytic shellfish poisoning. This disease spreads to humans when large concentrations of dinoflagellates (called a **red tide**) are eaten by mussels and clams, that in turn are eaten by humans.

Procedure

1. Observe *Peridinium* and *Ceratium* by making wet mount slides (if live cultures are available), or by using prepared slides.
2. Draw both organisms:

Peridinium   *Ceratium*

PHYLUM EUGLENOPHYTA

Euglenoids are flagellated, unicellular algae (Figure 28.2). They lack a cell wall, but have a semirigid plasma membrane called a pellicle. Although these organisms have chloroplasts, and are photosynthetic, some can also be heterotrophic or saprophytic. In the dark, they ingest organic matter through the **cytostome.** All euglenoids except one have a red **eyespot** at the anterior end. This organelle senses light and the organism can position itself, using its flagellum, in an appropriate direction.

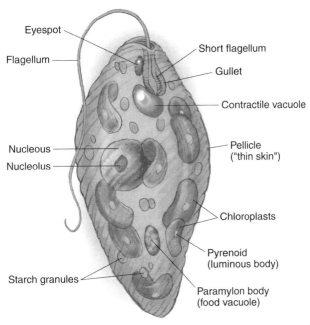

Figure 28.2
Euglena

Procedure

1. Observe *Euglena* by making a wet mount slide (if live cultures are available), or by using a prepared slide.
2. If you are using living material, add a drop of methylcellulose (proto-slow) in the preparation in order to slow the *Euglena*.
3. Draw *Euglena*.

PHYLUM BACILLARIOPHYTA

Diatoms are unicellular, filamentous, or colonial algae with complex cell walls composed of pectin and silica (Figure 28.3). The two parts of the wall fit together like the halves of a Petri dish. Diatoms contain the pigments chlorophyll a and c, and xanthophylls, which give these organisms a golden-brown color. Like the dinoflagelates, the diatoms are important primary producers in the ocean.

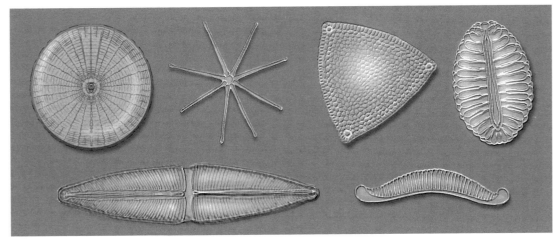

Figure 28.3
Diatoms

Diatoms store energy captured through photosynthesis in the form of oil. Much of the world's petroleum supply was formed from diatoms that lived over 300 million years ago. Also, large deposits of silica, from the walls of the diatoms, can accumulate in layers of a material that is called **diatomaceous earth.** Diatomaceous earth is used commercially in, for example, swimming pool filters.

Procedure

1. Make a wet mount slide using a small amount of diatomaceous earth in water.
2. Draw two types of shapes that you observe below:

Diatom   Diatom

3. Examine living material (if available) or a prepared slide of diatoms.
4. Draw two types of shapes that you observe below:

Diatom   Diatom

PHYLUM PHAEOPHYTA

The **brown algae** range in size from microscopic forms to kelps that can grow over 50 m in length (Figure 28.4). They have multicellular bodies and **holdfasts** that anchor them to rocks in shallow water along tidal coasts of oceans. Brown algae usually grow in temperate waters and get their color from the brown pigment, **fucoxanthin,** which masks the green chlorophyll that is also present. Some brown algae can grow at a rate of 20 cm per day, and can therefore be harvested for commercial use. **Alginic acid,** a thickener used in many foods such as ice cream and pudding, is extracted from the cell walls.

Procedure

Examine the examples of brown algae provided to you in lab, including *Fucus,* and draw them in the space below:

Fucus   Species

Figure C. Sargassum Sp.

Figure F. Fucus Sp.

Figure G. Laminaria Sp.

Figure H. Macrocystis Sp.

© Kendall/Hunt Publishing Company

Figure 28.4
Phylum Phaeophyta

Exercise 28, Part 1 Kingdom Protista—The Algae

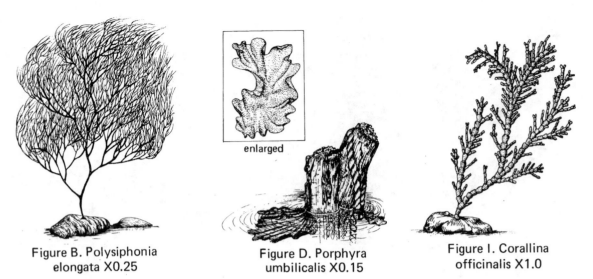

Figure 28.5
Phylum Rhodophyta

PHYLUM RHODOPHYTA

The **red algae** have delicately branched thalli and live at greater ocean depths than other algae (Figure 28.5). They obtain their color from the presence of the red **phycobilin** pigment. The thalli of a few red algae form crust-like coating consisting of calcium carbonate. These species, the **coralline algae,** help to build the bases of coral reefs. Agar and carrageenan, both commercially important products, are extracted from the red algae.

Procedure

Examine the examples of red algae provided to you in lab, and draw them in the spaces below:

Species: _____   Species: _____

PHYLUM CHLOROPHYTA

Green algae are the most diverse and familiar algae in both marine and freshwater habitats. They have cellulose cell walls, contain chlorophyll a and b, and store starch, as do plants. Therefore, the green algae are thought to have given rise to terrestrial plants. Green algae can be found in forms that are unicellular, filamentous, colonial, and multicellular (Figure 28.6).

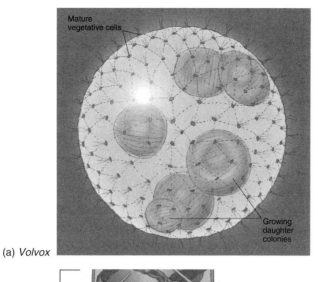
(a) *Volvox*

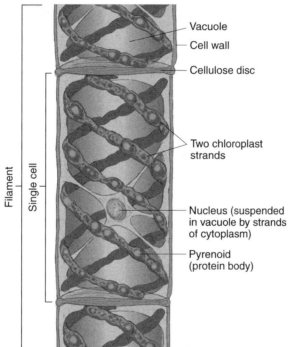
(b) *Spirogyra*

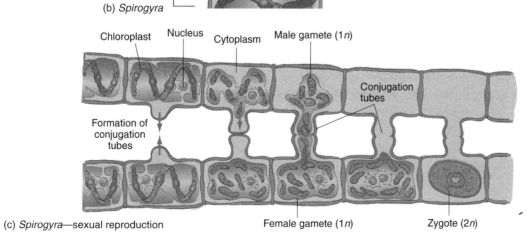
(c) *Spirogyra*—sexual reproduction

Figure 28.6
Phylum Chlorophyta

Unicellular Green Algae

1. Using living material, prepare a wet mount of *Chlamydomonas*. You may need to add a drop of methylcellulose to slow down the organisms.
2. Draw the organism below, labeling the nucleus, flagella, and the stigma, which appears as a reddish spot at the anterior end of the cell.

Chlamydomonas

3. Next, you will observe the process of **syngamy,** the pairing and fusion of haploid gametes to form diploid cells. The gametes (called isogametes because they appear to be identical) are referred to as + or –. The gametes unite to form the **zygote.**
4. Place drops of + and – gametes of *Chlamydomonas* next to each other on a slide. BE CAREFUL. Do not allow the drops to mix. Do not add a coverslip.
5. While you are looking through the microscope at the cells, mix the two drops using a toothpick.
6. Note the clumping of the gametes, and look for cells that are in pairs. Draw this below:

Chlamydomonas syngamy

Filamentous Green Algae

1. Using living material, make a wet mount slide of *Spirogyra* (Figure 28.6b).
2. Draw this organism below, labeling the chloroplast, pyrenoids (starch storage organs) in the chloroplast, the cell wall, and the nucleus.

Spirogyra

Colonial Green Algae

1. Using living material and prepared slides, observe *Volvox* (Figure 28.6a), a spherical colony consisting of thousands of cells, each bearing two flagella. During asexual reproduction, some cells move down INTO the sphere and form **daughter colonies,** which are eventually released.
2. Look for a variety of forms and find at least one with daughter colonies inside.
3. Draw two colonies, one containing daughter colonies. Label the nucleus, daughter colonies, cells, and flagella.

Volvox   *Volvox* (with daughter colonies)

Multicellular Green Algae

1. Using living material or preserved material, observe the green algae, *Ulva* (or sea lettuce).
2. Draw the *Ulva* below:

Ulva

QUESTIONS

*1. _____ Which algal group (brown, red, or green) includes "kelp"?

*2. _____ To which phylum do brown algae belong?

*3. _____ To which phylum do red algae belong?

*4. _____ To which phylum do green algae belong?

*5. _____ Which of the algae are probably ancestral to land plants?

*6. _____ Which of the algae are largest?

*7. _____ Which red alga is often confused with coral?

8. _____ Give an example of a filamentous green alga.

9. _____ Give an example of colonial green alga.

10. _____ Diatoms belong to the phylum.

*From *Experiencing Biology: A Laboratory Manual for Introductory Biology* 5th Edition by GRCC Biology 101 Staff. Copyright © 2000 by Kendall/Hunt Publishing Company. Reprinted by permission.

Name _____ Section _____ Date _____

Exercise 28, Part 2: Kingdom Protista—Protozoa and Slime Molds

PART II

OBJECTIVES

After completing this exercise, you should be able to:

1. Describe the characteristics specific to the protozoa and slime molds.
2. Identify selected examples of the protozoa and slime molds according to their mode of locomotion, nutrition, cellular organization, and means of reproduction.

INTRODUCTION

Protozoans are one-celled, eukaryotic organisms. They are predominantly **heterotrophs,** feeding, on bacteria and small particulate matter, and inhabit water and soil. Some are **parasitic,** living, on or in other organisms. Each phylum of protozoa is classified on the basis of its mode of locomotion.

In general, protozoans live in an area that has a large supply of water. Water can be accumulated or expelled by means of a **contractile vacuole.** Food can be transported directly across the plasma membrane. However, for species that possess a protective covering, food is ingested by means of specialized structures. Digestion takes place in **vacuoles,** and the elimination of waste occurs across the plasma membrane, or through an **anal pore.**

PHYLUM RHIZOPODA

The **amoebas** are found worldwide, in fresh and salt water, and in soil (Figure 28.7). Some forms of amoebas are parasitic. An example is *Entamoeba histolytica,* which causes amoebic dysentery. Amoebas move and feed by the use of **pseudopods** (Greek for "false foot"). Pseudopods are flowing projections of cytoplasm that extend and pull the amoeba forward, or encircle and engulf food. Amoebas reproduce asexually by a process called **fission.**

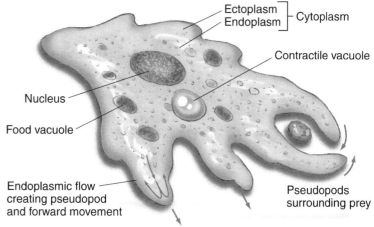

Figure 28.7
Amoeba

From *Biological Investigations* by Gayne Bablanian. Copyright © 2004 by Kendall/Hunt Publishing Company. Reprinted by permission.

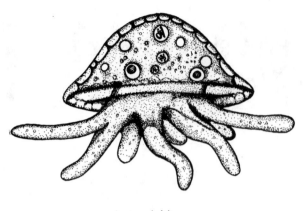 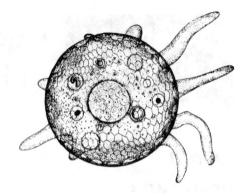

Lateral View Surface View

Figure 28.8
Phylum Foraminifera

© Kendall/Hunt Publishing Company

Procedure

1. Prepare a wet mount slide by removing a drop of water from the BOTTOM of the *Amoeba* culture using an eyedropper.
2. Under SCANNING POWER, look for granular, grayish, irregularly shaped "blobs." Then go to higher power to observe the organism.
3. Observe the *Amoeba* until you see it begin to move.
4. Describe this movement:
   _____
   _____
5. Add a drop of nutrient broth, and observe feeding behavior.
6. Draw the *Amoeba* below. Label the nucleus and pseudopods.

Amoeba

PHYLUM FORAMINIFERA

The **forams** are marine protists, and range in size from 20 micrometers to 3 centimeters (Figure 28.8). These "shelled amoebas" are surrounded by a covering (called a **test**) composed of organic materials and calcium carbonate. (The fossil remains of the tests are constituents of limestone formations, like the White Cliffs of Dover, in England.) The tests have pores through which the **podia** (feet) of the organism can emerge. Forams live in sand, are attached to other organisms, or are planktonic.

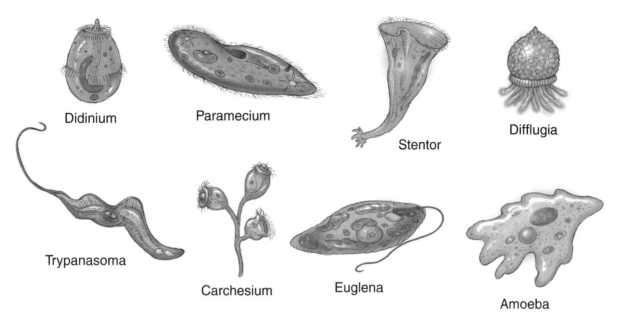

Figure 28.9
Phylum Zoomastigophora

Procedure

1. Examine a prepared slide of the tests of various forms of forams.
2. Draw two different types below:

Foram   Foram

PHYLUM ZOOMASTIGOPHORA

The **zoomastigotes,** or **flagellates,** are unicellular, variable in form, and have at least one flagellum (Figure 28.9). They include both free-living and parasitic types. The parasitic forms can live within various animals including humans. Some examples are *Giardia* (an intestinal parasite that causes Hiker's Diarrhea), *Trichomonas vaginalis* (a sexually transmitted protozoan), and *Trypanosoma* (causes sleeping sickness and is transmitted by the tse tse fly).

Procedure

1. Examine the prepared slides of *Trypanosoma*.
2. *Note:* You will also see red blood cells in this sample.
3. Draw a few blood cells and the *Trypanosoma* parasite. Label the blood cells, and the protozoan's nucleus and flagellum.

Trypanosoma

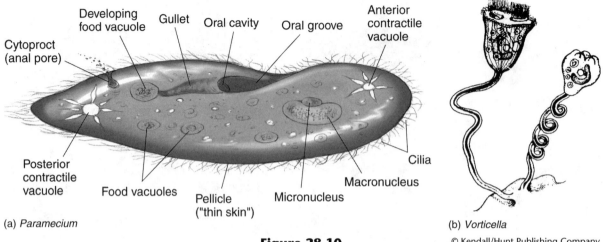

Figure 28.10

(a) *Paramecium*
(b) *Vorticella*

© Kendall/Hunt Publishing Company

PHYLUM CILIOPHORA

The **ciliates** move by means of cilia that are arranged either in spirals around the body of the organism or in longitudinal rows (Figure 28.10). The outer covering of the ciliates is called the **pellicle,** and is tough, but flexible. Because of this covering, food must be ingested through specialized structures like the **cytostome** found in *Paramecium*. The ciliates also possess two types of nuclei. The **micronuclei** are involved in reproduction and heredity, and the **macronuclei** in the control of the physiology of the cell.

A. *Paramecium*

Paramecium, a common freshwater protist, has cilia located around the cell, and in the **oral groove,** a long shallow depression which leads to the mouth or **cytostome.** Food entering the cytostome is deposited into a **food vacuole,** where the food is digested. Egestion of waste materials is through a part of the pellicle called the **anal pore.** Two **contractile vacuoles,** one located at each end of the cell, pump out excess water that has entered the cell. Asexual reproduction occurs by a process called **fission.** Sexual reproduction, if present, consists of the exchange of nuclei in a process called **conjugation.**

Procedure

1. Prepare a wet mount slide of living *Paramecium,* using a drop of methylcellulose to slow down the organisms.
2. How does the movement of *Paramecium* compare to that of *Amoeba*?
3. Draw *Paramecium* below. Using Figure 28.10 as a guide, label as many structures as you can.

Paramecium

B. *Vorticella*

Vorticella is not free-swimming, and consists of a **bell shaped body** attached to a long **contractile stalk.** The rim of the bell contains cilia arranged in three whorls. The cilia surround the oral groove that leads into the mouth (see Figure 28.10b).

Procedure

1. Prepare a wet mount slide if living material is available or observe *Vorticella* using a prepared slide.
2. Draw *Vorticella* below. Label the cilia, body, and contractile stalk.

Vorticella

3. In the living specimen, what is the function of the moving cilia that surround the mouth?

___

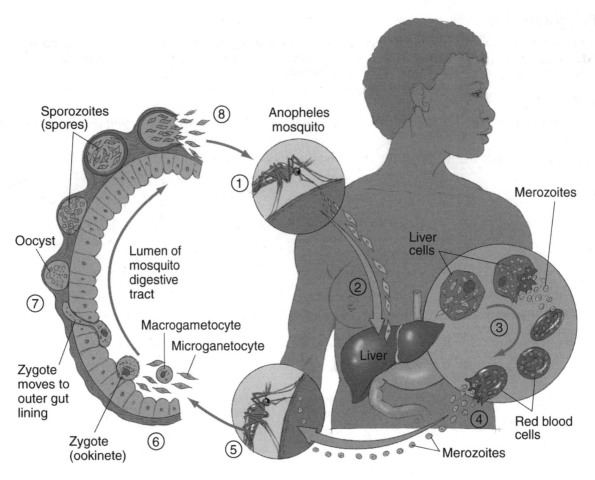

Figure 28.11
Life Cycle of *Plasmodium*

© Kendall/Hunt Publishing Company

PHYLUM SPOROZOA

The **sporozoans** are non-motile, spore-bearing parasites of animals. Their spores are small, infective bodies that are transmitted from host to host. The life cycle of these organisms is complex, and involves both asexual and sexual phases. The best known among the sporozoans is the parasite that causes **malaria,** *Plasmodium* (Figure 28.11).

Plasmodium is transmitted from human to human by means of mosquitoes of the genus *Anopheles*. These parasites first infect the cells in the liver. In the next stage, they enter the bloodstream, and infect and rupture the red blood cells, causing toxic substances to trigger cycles of fever and chills.

Procedure

1. Examine a slide of blood with parasites inside the red blood cells.
2. Draw an infected red blood cell along with several healthy red blood cells.

Plasmodium infected red blood cells.

SLIME MOLDS

Slime molds are divided into three groups, each with unique characteristics. The phylum **Acrasiomycota,** also known as the **cellular slime molds,** resemble amoebas. The phylum **Oomycota,** or the **water molds,** resemble fungi but possess cellulose in the cell wall, and have a different reproductive strategy. The phylum **Myxomycota** or **plasmodial slime molds,** exist as a mass of protoplasm with many nuclei. This mass of protoplasm is called a **plasmodium.** The entire plasmodium moves as a giant amoeba. Within the protoplasm, **cytoplasmic streaming** can be observed. The protoplasm moves so that oxygen and nutrients are distributed evenly.

Procedure

1. Observe a demonstration plate of *Physarium,* a plasmodial slime mold. You will need to use a dissecting microscope.
2. In which direction does cytoplasmic streaming occur?

    _____

    _____

QUESTIONS

1. _____ Do all protists have a cell membrane?

2. _____ May protists be enclosed in a cell wall?

3. _____ How many nuclei are found typically in *Paramecium?*

4. _____ What is the structure of motility in *Paramecium?*

5. _____ What is the structure of motility in the *Amoeba?*

6. _____ What semirigid structure gives many protists their nearly constant shape?

*7. _____ Protozoa are classified into groups according to their mode of _____.

8. _____ *Paramecium* moves by means of its_____.

9. _____ *Amoeba* moves by forming _____.

10. _____ *Plasmodium* causes the disease called _____.

11. _____ *Trypanosoma* causes the disease called _____.

---

*From *Experiencing Biology: A Laboratory Manual for Introductory Biology* 5th Edition by GRCC Biology 101 Staff. Copyright © 2000 by Kendall/Hunt Publishing Company. Reprinted by permission.

Name _____ Section _____ Date _____

Exercise 29 Kingdom Fungi

OBJECTIVES

After completing this exercise, you should be able to:

1. Describe the fungi in terms of their physical organization, mode of nutrition, means of reproduction, and classification.
2. Describe the characteristic features of lichens.

INTRODUCTION

The **fungi** are a large group of eukaryotic organisms that lack chlorophyll and utilize absorption as their means of nutrition. The unicellular fungi are called **yeasts. Molds** are multicellular filamentous organisms such as mildews, rusts, and smuts. **Fleshy fungi** are multicellular, and include mushrooms, puffballs, and coral fungi. All fungi are **heterotrophs** that require organic compounds made by other organisms for energy and carbon. The majority of fungi feed on dead organic matter. These are called **saprophytes.** By using **extracellular enzymes,** fungi are the primary decomposers of the hard parts of plants. Other fungi are **parasitic** and obtain their nutrients from a living host.

The fungal body **(thallus)** consists of long filaments of cells called **hyphae** (singular, **hypha**). Some phyla possess hyphae that contain crosswalls called **septa,** which divide the hyphae into distinct cells with one nucleus. Other hyphae contain no septa and appear as long, continuous cells with many nuclei. These hyphae are called **coenocytic.** Hyphae grow and intertwine to form a mass called a **mycelium.** The **vegetative mycelium** is the portion of mycelium involved in nourishment. The portion concerned with reproduction is called the **aerial mycelium.**

Reproduction in fungi can be sexual or asexual. Yeasts reproduce by a process called **budding** (Figure 29.3). The parent cell forms a bud on its outer surface. As the bud elongates, the parent's cell's nucleus divides, and one nucleus migrates into the bud. The bud will eventually break away.

Another reproductive strategy is **spore** (sporangia) formation. Spores are formed from the aerial mycelium in a number of ways. **Asexual spores** are formed from the aerial mycelium of a single individual. After germination of the spore, the new organism is genetically identical to the parent. The different types of asexual spores are shown in Figure 29.1. **Sexual spores** are formed from the union of nuclei from two genetically different individuals. An example of sexual spore formation is shown in Figure 29.2. Fungi are classified into four phyla, depending upon the type of sexual spore that is present. The phyla are the **Zygomycota, Ascomycota, Basidiomycota,** and the **Deuteromycota** (or *Fungi Imperfecti*). The Deuteromycota are "imperfect" because no sexual spores have yet been identified or found for individuals in this grouping.

From *Biological Investigations* by Gayne Bablanian. Copyright © 2004 by Kendall/Hunt Publishing Company. Reprinted by permission.

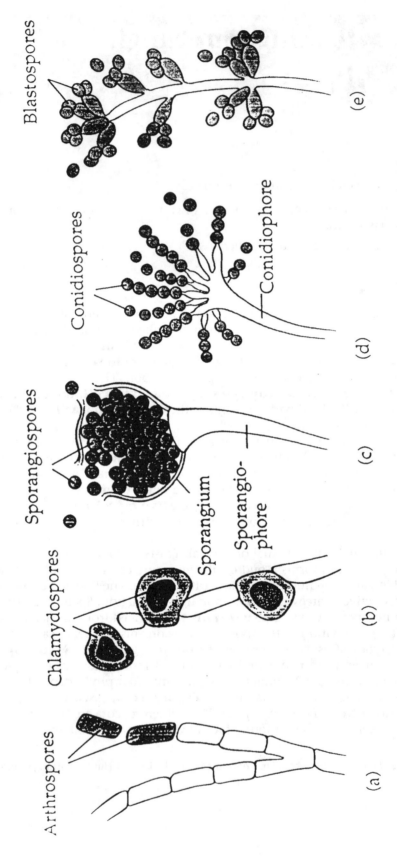

Figure 29.1

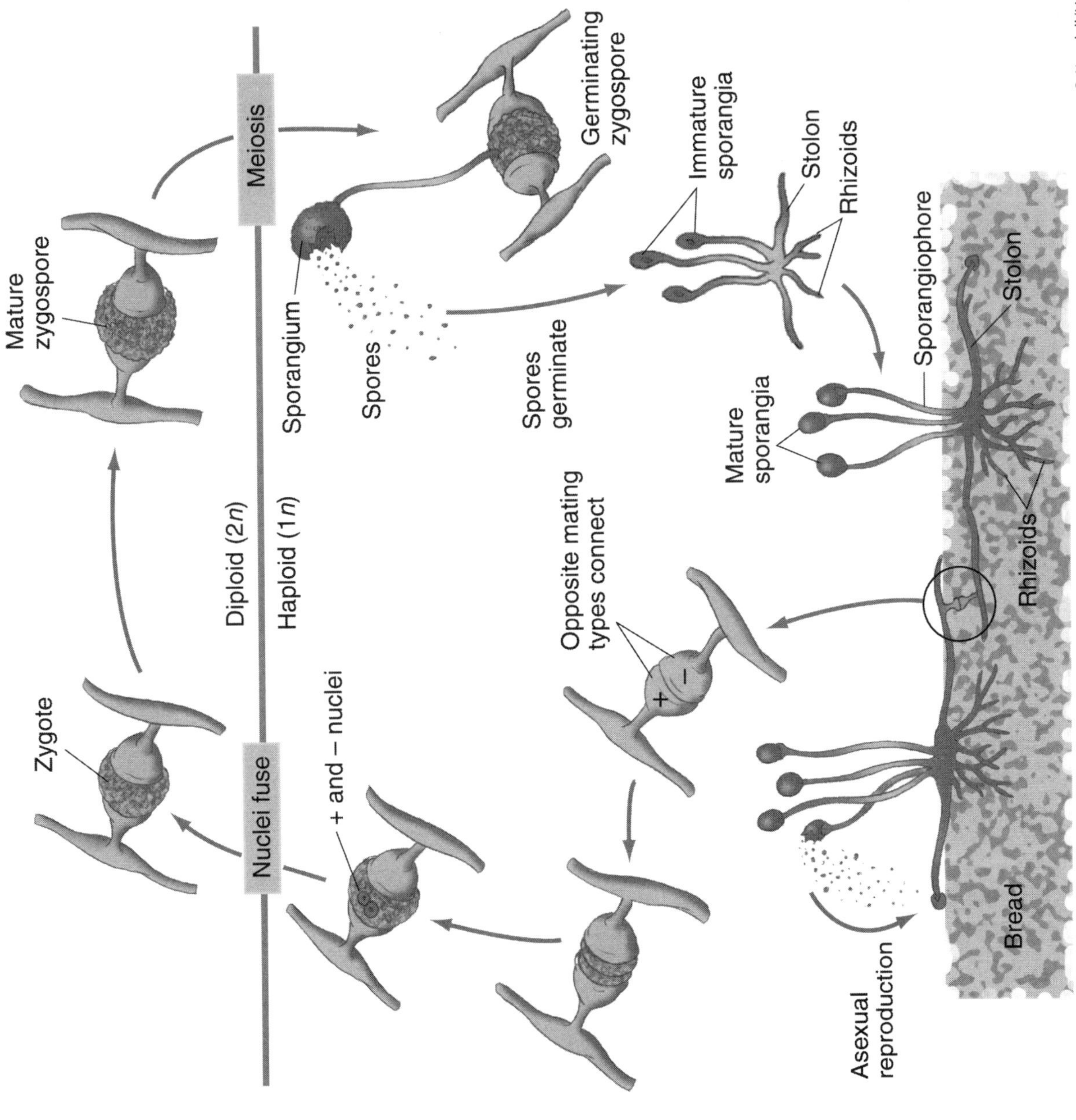
Zygote
Nuclei fuse
+ and – nuclei
Diploid (2n)
Haploid (1n)
Mature zygospore
Meiosis
Germinating zygospore
Sporangium
Spores
Spores germinate
Immature sporangia
Stolon
Rhizoids
Opposite mating types connect
+
–
Mature sporangia
Sporangiophore
Stolon
Rhizoids
Bread
Asexual reproduction
+ mating type
– mating type

Figure 29.2

PHYLUM ZYGOMYCOTA

The **zygomycota** include the common bread molds, water molds, potato blight, and others. They are saprophytic and have coenocytic hyphae. The example you will be studying is *Rhizopus nigricans*, the black bread mold. The asexual spores of *Rhizopus* are **sporangiospores** and are located on asexual reproductive structures called sporangiophores (Figure 29.2). The spores inside the sporangium have a dark color, hence the descriptive name given to *Rhizopus*. The sexual spores are **zygospores**.

The mycelium is differentiated into several specialized structures. The **rhizoids** are branches that extend into the substrate and help in anchoring the fungus, and in absorption. **Stolons** connect different sections of hyphae. You should be able to observe all these structures in lab.

Procedure

1. Obtain *Rhizopus* that has been grown in a Petri dish. The dish is sealed. DO NOT open it.
2. Using a dissecting microscope observe the hyphal filaments making up the mycelium.
3. List the structures that you can identify:

   _____

   _____

4. Obtain a prepared slide of *Rhizopus*. Find a section from which you can draw the major structures of the fungus. Label each structure.

Asexual Reproduction   Sexual Reproduction

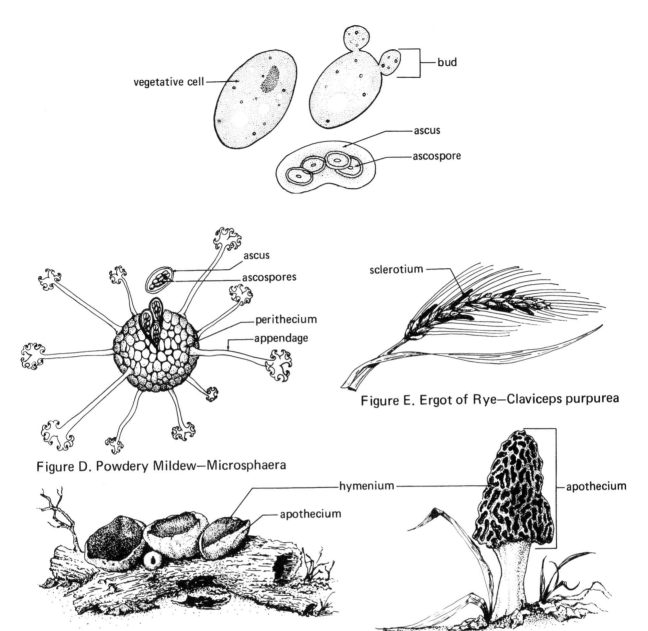

Figure 29.3

© Kendall/Hunt Publishing Company

PHYLUM ASCOMYCOTA

The **Ascomycota,** or sac fungi, include the yeasts, morels, truffles, and some molds. Yeasts are used commercially for producing baked products, and beer and wine. Morels and truffles are edible fungi. However, there are also harmful parasitic forms such as Dutch Elm Disease and Chestnut Blight. Ascomycetes are called "sac fungi" because their sexual spores are **ascospores,** produced in a sac, or **ascus.** Their asexual spores are usually **conidiospores** (Figures 29.1 and 29.3).

Yeasts

1. Prepare a wet mount slide from a stick culture of *Saccharomyces*. Add a drop of either cotton blue or methylene blue stain.
2. Draw individual yeast cells below:

Saccharomyces

3. Obtain the fungi *Aspergillus* and *Penicillium* that have been grown in a Petri dish. DO NOT open the sealed dish.
4. Using a dissecting microscope, observe the hyphal filaments and mycelium.
5. List the structures that you can identify:

_____

_____

PHYLUM BASDIOMYCOTA

The **basidiomycota** is a diverse group that includes the mushrooms, shelf fungi, puffballs, and parasitic forms like the rusts and smuts (Figure 29.4). The common name, club fungi, is derived from the shape of the **basidium,** which bears the sexual **basidiospores.** Their asexual spores are sometimes **conidiospores.**

Mushrooms are the most familiar fungi (Figure 29.5). The body of the mushroom is divided into the cap **(pileus)** which is situated on a stalk **(stipe).** On the undersurface of the cap are the gills where the spores can be found. When the mushroom is young, a covering **(veil)** extends from the cap to the stalk. In the mature mushroom, the remains of the veil can be seen as ring-like **annulus.**

Procedure

1. Obtain a fresh mushroom, *Agaricus campestris* (the white mushroom).
2. Draw the external features, and label the structures of the mushroom.

Agaricus campestris

3. Remove the cap, and with a sharp razor blade, or scalpel, cut a THIN section of the gill. (If you need help with this, call over the instructor.)
4. Prepare a wet mount slide.
5. You should be able to observe a continuous **hymenial layer** along the entire gill margin. Projecting from the **hymenium** are the **basidia** that contain the **basidiospores.** A **sterigma** connects the spores to the basidium.
6. Draw this below and label:

Section of the gill

Figure 29.4

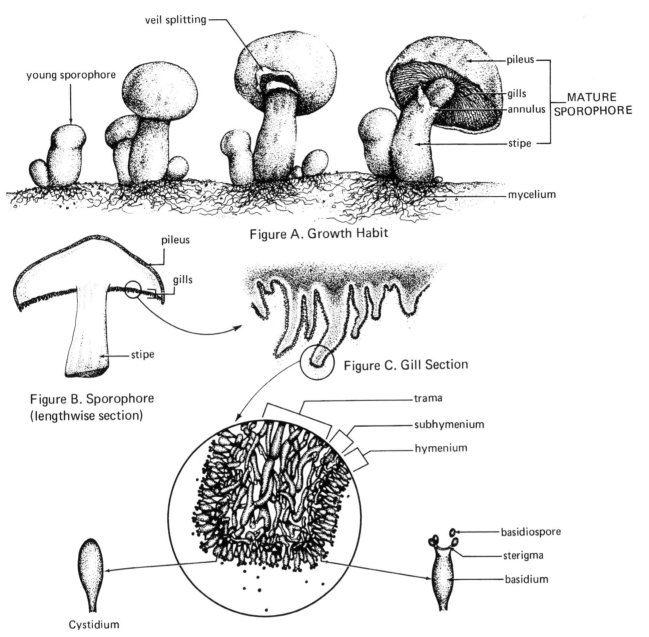

Figure 29.5

Exercise 29 Kingdom Fungi

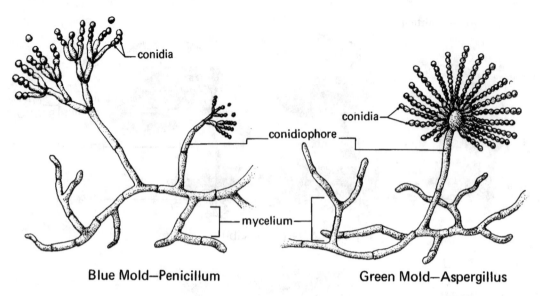

Figure 29.6

PHYLUM DEUTEROMYCOTA

The deuteromycota **(Fungi Imperfecti)** are the fungi that have no sexual stage. Many reproduce by the formation of **conidia** that are arranged in chains at the end of a **conidiophore** (Figure 29.6).

Procedure

1. Obtain a prepared slide of Penicillium and of Aspergillus.
2. Observe the formation of conidia.
3. Draw each below, labeling the hyphae, mycelium, conidia, and conidiophores.

Penicillium   *Aspergillus*

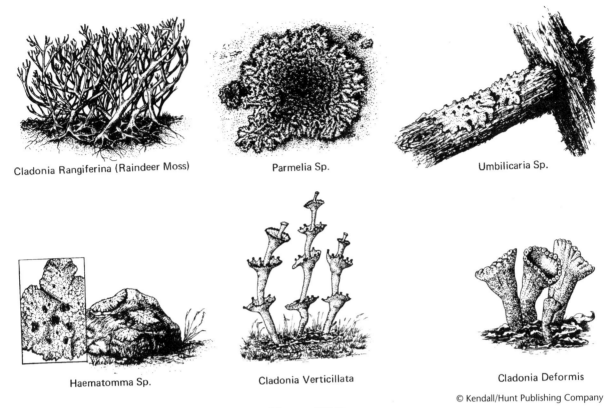

Figure 29.7
Representative Lichens

LICHENS

A lichen is a combination of a **fungus** and a **green alga** (or cyanobacterium). These organisms live together in a **symbiotic relationship.** Specifically, it is a **mutualistic** relationship, in which both organisms benefit. The fungus absorbs water and minerals, and retains moisture while living off of the sugars produced photosynthetically by the alga. The alga is supplied with moisture and is protected by the fungus.

There are about 25,000 species of lichens that inhabit a range of diverse habitats. They are some of the first organisms that can inhabit newly exposed rocks or soil. Lichens can withstand drought and extremes in temperature, but are sensitive and will die from pollution.

Lichens are grouped into three categories based on their morphology (Figure 29.7). **Crustose** lichens form crusts and grow flush with a hard substrate like tree bark, soil, or rock. **Foliose** lichens are leaf-like. **Fruticose** lichens have finger-like projections, or stalks.

Procedure

1. Examine a prepared slide of a cross section of lichen.
2. Draw the lichen below and distinguish between the fungus and the alga.

Lichen section

3. Examine the display of lichens in the lab. Note the morphological type and draw each below:

        Crustose   Foliose   Fruticose

QUESTIONS

1. _____ What is the nutritive of most fungi?

2. _____ To which kingdom do fungi belong?

3. _____ What is the name given to organisms which feed on dead or decaying organic matter?

4. _____ Are any fungi parasites?

5. _____ What is the basic reproductive structure in fungi?

6. _____ Which hypha function similar to the "runners" of strawberries?

7. _____ What are the rootlike structures of fungi?

8. _____ What kind of spores are produced by the Phylum Ascomycetes?

9. _____ What kind of spores arc produced by the Phylum Basidiomycetes?

*10. _____ Since fungi cannot make their own food, they are _____

*11. _____ and their mode of nutrition is by _____

*12. _____ The basic filaments of fungal growth are called _____, and _____

*13. _____ A mass of these filaments is called a _____

*14. _____ A lichen is an organism composed of a(n) _____

*15. _____ and a(n) living symbiotically in _____

*16. _____ a relationship called _____

*From *Experiencing Biology: A Laboratory Manual for Introductory Biology* 5th Edition by GRCC Biology 101 Staff. Copyright © 2000 by Kendall/Hunt Publishing Company. Reprinted by permission.

*17. When you leave the top off of a jar of food, or leave bread uncovered, it is likely to become "moldy." Where might the mold have come from?

*18. Many people are "allergic to fungi." They cough and sneeze and sniffle. Couldn't they avoid their suffering if they just would not eat mushrooms? Explain.

*19. During April each year, many people come to Michigan to search for the delicious mushroom called the morel. The best mushroom-collecting season is generally during the damp periods that occur in spring and autumn. Why do you suppose that is?

*20. Many fungi produce antibiotics, such as penicillin, which are valuable in medicine. Of what value might the antibiotics be to the fungi that produce them?

*From *Experiencing Biology: A Laboratory Manual for Introductory Biology* 5th Edition by GRCC Biology 101 Staff. Copyright © 2000 by Kendall/Hunt Publishing Company. Reprinted by permission.

Name _____ Section _____ Date _____

Exercise 30: Cultivation of Mushrooms

Primitive records picture the Egyptian Pharaohs eating and enjoying mushrooms. True cultivation of mushrooms began in the 18th century in France, where the mushrooms were grown in the dark cold caves of the country. The first mushrooms grown commercially in the United States are believed to have been grown in Long Island, New York, in 1880. Today, the cultivation of mushrooms has become much more technical. Within the last 25 years, new processing techniques have made growing mushrooms a finely tuned agricultural science.

In your Pulpit Rock Mushroom Kit you will find a large bag of compost containing spawn that is already grown. Before you receive your compost, it has gone through the following process:

Day 1–21: Compost is made from horse manure with small amounts of nitrogen, gypsum, and water added. This is composted over a period of 21 days.

Day 22: The compost is put on shelves in a closed room and cooked 7 to 10 days at 140°F to complete the sterilization process.

Day 33: The temperature is dropped to 75°F and the compost is inoculated with a spawn of the Agarus type, obtained from mushrooms through a laboratory process. When the compost becomes a gray moldy mass it is bagged, and added to your Pulpit Rock Mushroom Kit along with a bag of Canadian Deep Peat Moss with lime added as a PH adjustment and sent to you.

GETTING STARTED

- Open your box and pull the plastic liner to the top edge of the box.
- Level the compost making sure to get it evenly into the corners.
- In a separate container, thoroughly moisten all of the contents of the small bag of casing material with at least 3 or 4 cups of water, then spread it over the surface of the compost. **Do not press down.** (Try the moisture test to make sure the casing is thoroughly moistened.)
- Refold the liner over the surface and close the lid.
- Place the box away from direct sunlight in an area where the temperature is between 70°F and 75°F.

GROWING PERIOD

- After 10–15 days open the box to inspect the growing surface. If a gray threadlike mold is spreading over the top of most of the soil, cut the plastic liner away from the growing surface flush with the edge of the box. At this stage mist the surface heavily. *(If mold is not covering **most** of the surface, close up the liner and check again in 3–5 days.)*
- Place the flaps on the top of the box in an upright position and place cardboard or paper on top to prevent draft.
- Store in a cool place 55°–65°F. Over the next 7–14 days, mushroom growth appears as "pin heads" (tiny immature mushrooms). At this stage, keep the growing surface slightly damp. **Do not over water.**

  (Occasional light spraying with a plant mister is ideal at this stage.)

- Follow the watering procedure until mushrooms mature.

THE HARVEST

- You can harvest your crop of mushrooms during any of the three stages of growth: The amount of mushrooms that are produced is governed by the amount of food in the compost and other factors. If your environment is ideal, you may get a full crop on the first harvest; if not, you may have more than one harvest or several smaller crops.

    Buttons—small unopened mushrooms are very tender and light in taste.

    Fancies—slightly larger and more flavorful mushrooms.

    Flats—largest, most flavorful and nutritious mushrooms. The stage of growth when the head "flattens out" and the spore producing gills are exposed on the underside of the mushroom cap.

- When picking mushrooms, twist gently and lift away. Remove all roots or stumps. If the mushrooms are in clumps, cut off matured mushrooms with a sharp knife.

- With proper care, your kit will continue to produce mushrooms for 3–5 weeks.

Exercise 31: Kingdom Plantae

The members of this plant kingdom are all multicellular, eukaryotic organisms which are photosynthetic. Plant taxonomists used the word "Division" inside of phylum for the major plant groups. Current classification schemes list 12 divisions. For simplicity, we will consider two larger groups, the nonvascular and the vascular plants.

NONVASCULAR PLANTS

The nonvascular plants include the mosses, liverworts, and hornwarts, which are collectively known as the **bryophytes.** The bryophytes require water for reproduction and lack the woody tissue required for support of tall plants on land. Subsequently, bryophytes are small in height and are found in damp places (Figure 31.1).

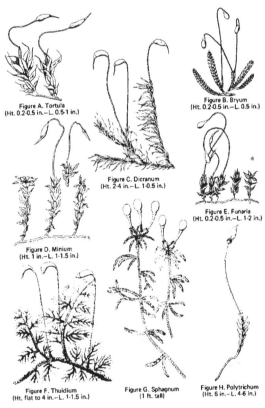

© Kendall/Hunt Publishing Company

Figure 31.1
Some Common Mosses

Portions of this exercise are from *Principles of Biology Laboratory Manual, 4th Edition* by H. B. Reid. Copyright © 2001 by Kendall/Hunt Publishing Company. Reprinted by permission.

Portions of this exercise are from *Introductory Biology 103, Laboratory Manual* 3rd Edition by P. Shields and H. B. Cressey. Copyright © 2002 by Kendall/Hunt Publishing Company. Reprinted by permission.

Portions of this exercise are from *Principles of Biology 112, Lab Manual* by Dept. of Biology, Ball State University. Copyright © 1996 by Kendall/Hunt Publishing Company. Reprinted by permission.

From *Concepts of Biology Laboratory Manual* 3rd edition by Boise State University Department of Biology. Copyright © 2002 by Kendall/Hunt Publishing Company. Reprinted by permission.

VASCULAR PLANTS

The adaptation to land led to the development in plants of an efficient vascular system made up of xylem and phloem. We generally further divide the vascularized plants based on the use of seeds for reproduction.

Seedless Plants

While there are many examples of seedless vascular plants, the most common example is the fern (Figure 31.2). Most ferns have leaves that are called fronds. Ferns have a specialized life cycle that includes both a diploid and haploid stage.

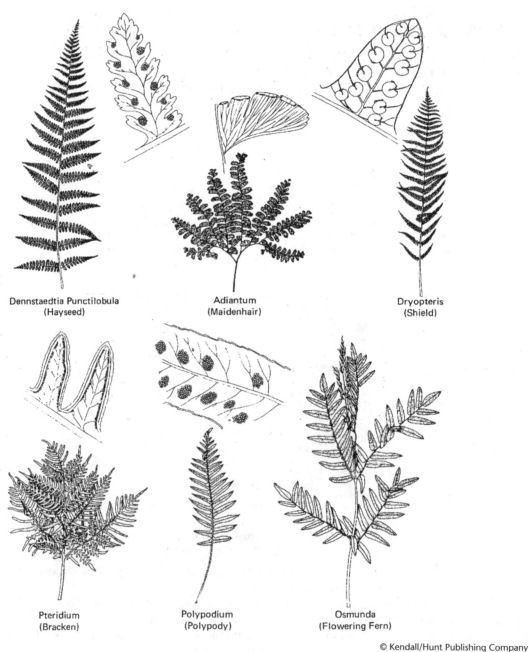

Figure 31.2
Some Common Ferns

Seed Plants

There are two major groups:

1. **Gymnosperms** include the conifers, cycads, ginkos, and gnetae. By far, the largest group is the conifers, which include pines, firs, and sequoias. These plants are flowerless and the gametophyte develops in specialized structures called cones (Figure 31.3).

2. **Angiosperms** include all flowering plants, and are distinguished by the presence of flowers and fruit. As can be anticipated, there are many subdivisions within this category, and over 235,000 species are known, with many more yet to be defined.

Figure A. Ginkgo biloba

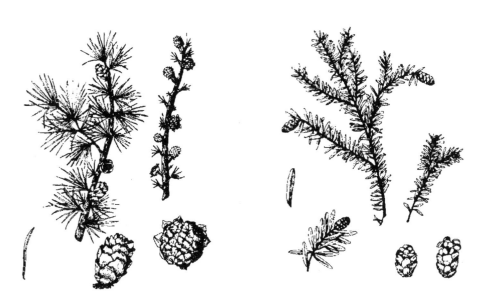

Figure B. Larch, Tamarack (Larix americana)

Figure C. Hemlock (Tsuga canadensis)

© Kendall/Hunt Publishing Company

Figure 31.3
Some Common Gymnosperms

FLOWER STRUCTURE: COMPLETE VERSUS INCOMPLETE FLOWERS

The four basic flower parts occur at the tip of a stalk called a peduncle (Figure 31.4). At its tip, the peduncle is enlarged to form the **receptacle** from which the four basic flower parts (referred to as **whorls**) arise. The outermost whorl of flower parts is known as the **calyx** and is composed of the **sepals.** Each sepal is a leaf-like structure, usually green, that covers and protects the flowers when it is still a bud. As some flowers mature the sepals change from green to the color of the petals (usually) and are often difficult to distinguish from the petals unless examined closely.

The next whorl of leaf-like structures is the **corolla,** which is composed of the **petals.** The petals are usually some color other than green, and generally function to attract birds or insects to aid in pollination. Flowers with large white petals and a strong fragrance are often pollinated by night flying insects (moths) or mammals (bats). Plants pollinated by the wind often lack petals completely.

In the center of the corolla is the **carpel** (also known as the pistil), the seed-bearing or female part of the flower. The carpel consists of three parts. At the base, attached to the receptacle, is a swollen structure called the **ovary.** Within the cavities of the ovary (locula) are the egg-bearing structures called **ovules**. A long, slender, neck-line extension from the tip of the ovary is called the **style.** At the tip of the style is a sticky swelling called the **stigma.** Pollen, blown by the wind or carried by an animal, gets stuck on the stigma and the **pollen tube** grows down the style to reach the ovules within the ovary.

The male portion of a flower consists of a variable number of **stamens.** Each stamen consists of a slender stalk-like **filament** with a sac-like structure at its tip called the **anther.** The stamens are attached to the receptacle around the base of the carpel. The anthers contain the **pollen grains,** the male gametophytes.

During the 150 million years of evolution of flowers, it is thought that parts became fewer and definite (3, 4, or 5) in number. In many flowers, parts are fused or lost, or clustered together to form an influorescence. Much taxonomic classification of angiosperms is based on floral characteristics.

The **dicot flower** usually has flower parts arranged in fours or fives. Four (or five) sepals, four petals, four stamens, and a carpel with four locules or cavities containing one or more ovules each. The **monocot flower** typically has flower parts arranged in threes, or multiples of threes.

A **complete flower** has all four parts: sepals, petals, stamens, and carpel. If one or more of the whorls are absent, it is an **incomplete flower.**
Use the information above to identify the parts of the flowers.

FRUIT TYPES

Study the fresh and preserved fruits available and compare them with the flowers and fruits shown in Figures 31.4 through 31.8. Classify these fruits using the scheme described below. Try to find one example for each of the types of fruits mentioned in the classification. Consult the wall charts and your textbook for types for which no examples are available.

The following broad scheme of classification of fruits, which is based on simple, easily observable but superficial characteristics, is artificial; that is, different plants that happen to have the same type of fruit according to this classification are not necessarily related closely.

1. Simple fruits—derived from the ovary of a single pistil. There are two types of simple fruits, fleshy and dry.
    1. Fleshy fruits—these are soft, juicy, or meaty. They are further classified as:
        a. Berry—with one or more carpels and seeds, and with an entirely fleshy or soft interior. This skin is thin, and in many cases can be peeled off.
        b. Pepo—a berry with a hard rind that cannot be peeled off.
        c. Hesperidium—a berry with a leathery rind that peels off.
        d. Pome—with several carpels; the fleshy part is derived from the receptacle and the true ovary forms a "core."
        e. Drupe—with a single carpel; single seed enclosed in a hard pit or stone.

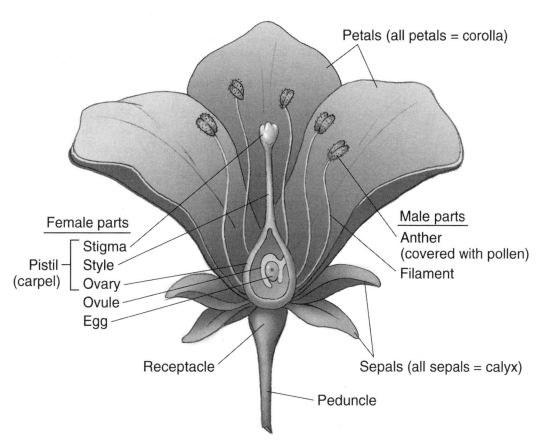

Figure 31.4
Generalized Structure of a Dicot Flower

2. Dry fruits—the wall of the mature ovary is dry; the fruit is without soft "flesh." Dry fruits are classified according to whether or not they split open when ripe.

   a. Dehiscent fruits—these split open when ripe. They are further classified as:

   1. Follicle—with one carpel; split along one suture.
   2. Legume—with one carpel; split along two sutures.
   3. Capsule—with many carpels; split in many different ways.

   b. Indehiscent fruits—these are dry fruits which do not split open when ripe. They include

   1. Achene—small, single-seeded fruit with a relatively thin wall, and with the seed attached to the wall only at one point.
   2. Grain—single-seeded; with the fruit wall completely fused with the seed coat.
   3. Samara—one or two carpels; with fruit wall forming winglike expansions.
   4. Nut—two or more carpels; fruit wall hard or stony.

2. Aggregate Fruits—derived from a single flower with many separate pistils.
3. Multiple Fruits—derived from many flowers clustered together so closely that their ovaries, together with many accessory parts, fuse together forming a single composite fruit.
4. Parthenocarpic Fruits—seedless fruits are said to be parthenocarpic. Many varieties of plants, especially cultivated ones, produce seedless fruits, in which the ovary ripens without pollination or fertilization. Seedlessness can be natural, or it can be induced by treatment with hormones.

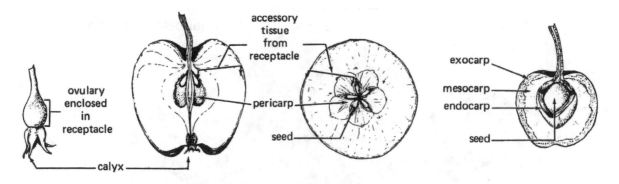
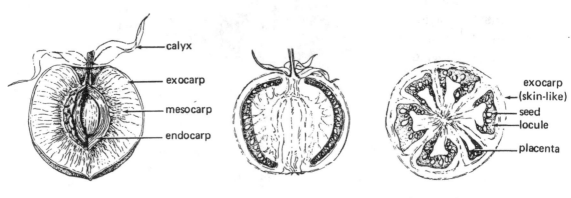
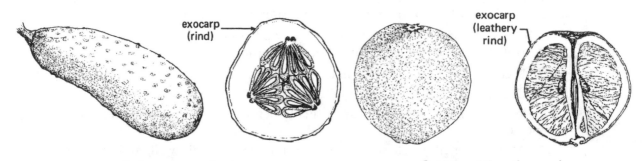

Figure 31.5
Simple Fleshy Fruits

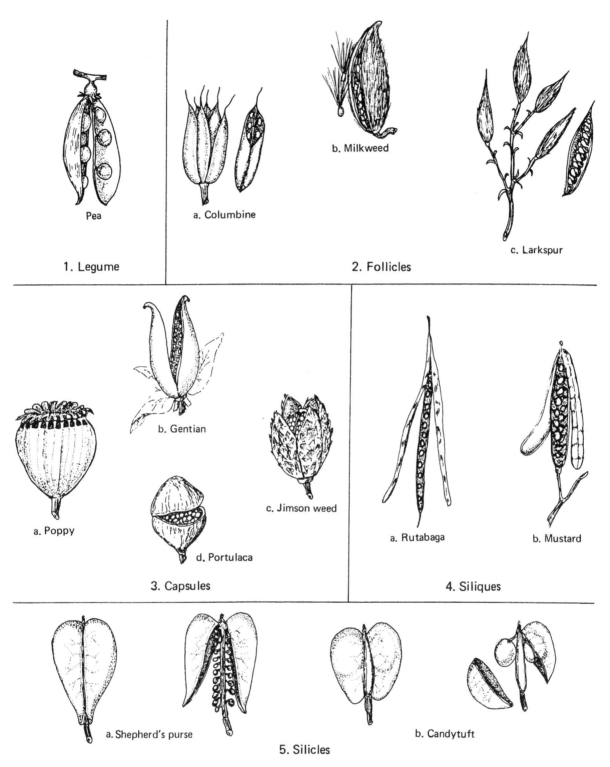

Figure 31.6
Simple, Dry Dehiscent Fruits

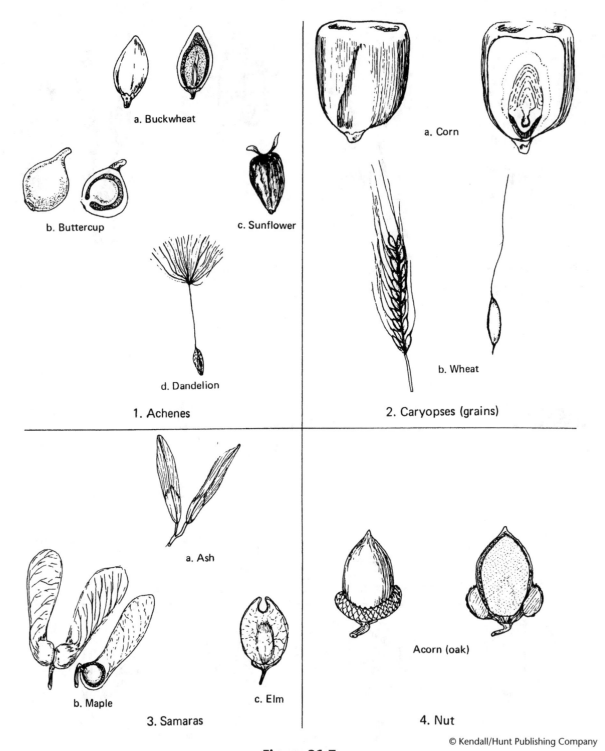

Figure 31.7
Simple, Dry Indehiscent Fruits

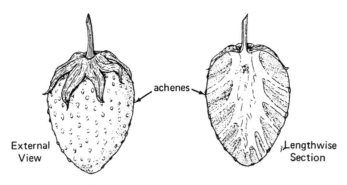

1. Aggregate Fruit—Strawberry

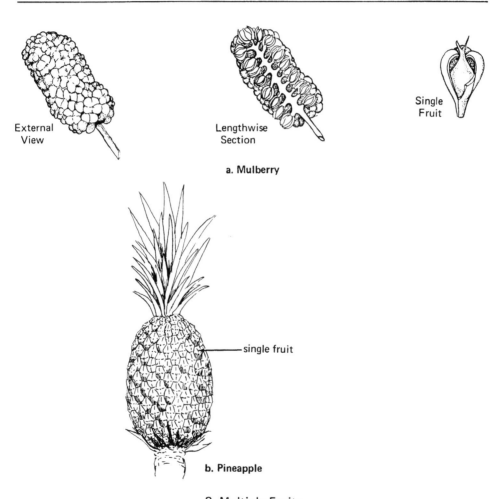

a. Mulberry

b. Pineapple

2. Multiple Fruits

© Kendall/Hunt Publishing Company

Figure 31.8
Aggregate and Multiple Fruits

Name _____ Section _____ Date _____

Exercise 32 Botanical Field Project
Developed by Dr. Flisser—http://dflisser.pageout.net

Keep track of this guide—All details are important as you proceed through the project.
  This is a group activity and emphasis is placed on assisting one another. The **herbarium** specimens which you submit, and the narrative you write, however, will be *individualized*—each student will submit his/her own completed project.

1. Project makeups are treated in the same manner as exam makeups:
2. With **valid documentation,** an **alternate** project will be given.

OBJECTIVES
We are interested in two goals as the class carries out this field project:

1. Familiarization with the taxonomy and identification of **local flora.**
2. Conducting a **botanical survey** of the flora present at a particular ***time and place.***

Local Flora
The plants of the Camden County College campus may be divided into two groups based on their location:

A. Horticultural/Ornamental Plants

These were specifically planted on campus for various esthetic reasons. Horticultural varieties are generally valued for their flowers, fruits, green foliage, or their fall colors.
  The horticultural plants are found throughout the "open" region of the campus, in and among the buildings. The ornamental plants include trees, shrubs, and annual plants in flower beds.

B. Naturally Occurring Plants

These comprise the indigenous flora of our region of the Eastern Deciduous Forest Biome. Such plants thrived well before the college was built here.
  The natural flora is found primarily in the 300+ acres of woodland at the south end of campus. This flora is divided according to three layers of *vertical stratification* (see page 199).

Botanical Survey
Botanists are interested in producing an inventory, or a record, of the flora of specific regions every few years, in order to investigate the changes which may appear.
  The botanical inventory may reveal that surprising plants, more common in distant ecosystems, occasionally appear unexpectedly. At the same time, plants which had appeared in moderate numbers in previous surveys, may drop out of view.
  Such changes in the floristic composition are usually indications of subtle environmental fluctuations which may need examination.

Plant Migration
Changes in plant populations that are revealed in periodic botanical surveys suggest that plants actually migrate. Their travel, however, occurs in a more subtle manner than animal migration.

There are several results of plant migration:

Establishment of New Populations

1. Seeds, fruits, fragments of stem and root are carried by insects, birds, mammals, wind, and water to regions where the plant may not have grown before.
2. If the plant is successful in its new habitat, a population may become established.

Loss of Populations

1. On the other hand, in New Jersey, the most densely populated state of all 50, we are faced with an accelerated case of **habitat destruction.**
2. An important axiom among ecologists is that, *in order to save a species, one must preserve its habitat and environment.*
3. Conversely, habitat destruction leads to species loss, and their disappearance from subsequent surveys.

Importance of Location and Date

Plants carry out different aspects of their life cycles at particular times of the year.
  Among the campus examples:

1. Apples and cherries flower in spring, and produce fruits that mature in late summer.
2. Goldenrod flowers in late summer, and fruits appear in early fall.

And, of course, particular plants are known to inhabit specific regions. This distinguishes tropical flora from temperate plants. Even within a region, habitat specificity separates plants adapted to dry, moist, salty, sunny, and shaded regions.
  The point is that, in conducting any botanical survey, it is important to record the exact date and exact location at which plants were collected. Since the botanical survey is carried out for future reference, carefully prepared data must be provided for those who will work with this in years to come.

CAMDEN COUNTY COLLEGE CAMPUS

A trip through the wooded section of the college campus illustrates one small fraction of the variety of flora comprising the Kingdom Plantae. One objective of this exercise will be to distinguish among the various categories of plants.
  There are two levels on which to view the wooded section of our campus, or any region:

1. **Ecological**—This refers to the habitat characteristics and community associations of the plants occurring over the general region. What are the interactions among different organisms with the habitat and among one another?
2. **Organismal**—This level of study refers to the individual trees and plants. What are the differences among them? How is each plant unique?

Ecological Overview

The natural flora of Camden County College is a part of the **eastern deciduous forest,** a biome which includes several distinct plant communities:

1. The **hardwood** forest includes familiar trees such as Oak and Hickory in dry areas, and Beech, Birch, Maple, and others in the regions with greater moisture. These trees are *deciduous,* and prefer well-drained soil.

2. The ***coniferous*** region is composed of evergreen needle-leaved trees such as Pines, Spruce, Hemlock, and Juniper, and prefer a more acidic soil. Because of the acidic nature of the soil, there are generally few understory shrubs beneath a coniferous forest.

Vertical Stratification occurs in the hardwood forest, as the community units are divided among three physical levels:

1. **Canopy**—This level is characterized by trees whose crowns overlap and form an unbroken covering. In our region, the associations differ depending on moisture availability:

   Dry sites: Oak, Hickory

   Increased moisture: Beech, Birch, Maple

2. **Understory**—Shade-tolerant trees and shrubs, including young individuals of the canopy species. These survive well beneath the canopy. Periodically, a taller, canopy tree is lost to a lightning strike, or natural death. When this occurs, understory trees take advantage of the opening and grow to large size in the presence of increased light. This is especially true of the understory members which are juvenile versions of the dominant species.

3. **Ground Level**—These plants are generally herbaceous (non-woody) and include wildflowers, ferns, mosses, and grasses. The presence of particular plants at ground level depends on various levels of light filtering down through the trees, and on the differing qualities of soil.

Organismal Overview

An organism is an individual living thing. The organismal view of the Camden County College campus requires observation of the individual plant and animal inhabitants.

For all organisms, the "bottom line" characteristic in biological classifications generally is the reproductive structure. Therefore, when studying the plant categories listed below, you are asked to take note of their reproductive features.

Kingdom Plantae

Flowering Plants (Angiosperms)

Any plant that produces a flower and fruit as its reproductive structures. This includes the dominant canopy trees, members of the understory, and ground level wildflowers. These are the dominant plants on Earth today.

Conifers (Gymnosperms)

Forest trees such as Pines, Spruce, Fir, Hemlock, having needle-leaves, and producing cones as reproductive structures. They are often called evergreen, although those such as Bald Cypress (ornamentally planted on campus) are actually deciduous.

Ferns

Herbaceous plants with feather-like compound leaves, often growing in shade. Reproduction takes place via spore production. The spores often occur along the lower surface of the leaves.

Fern Allies or Club "Mosses"

Small, trailing, ground-level herbaceous plants, which are neither ferns nor mosses. They do have a reproductive system reminiscent of ferns, and produce spores which germinate as a new generation. A large population of *Tree Club-Moss* will be found in our woods. To confuse the issue further, they are sometimes given names such as *Ground Cedar* or *Running Pine*, which are the names of Conifers.

Mosses and Liverworts

Small, nonvascular, green plants, growing in extensive mats, often in shady areas. Sometimes these habitats are moist, though not always.

Kingdom Protista

Algae

Aquatic, nonvascular, photosynthetic organisms, which are actually members of the Kingdom Protista. They may grow on the water surface or on moist rock and other surfaces. These may be green, brown, red, or gold in color.

Kingdom Fungi

Fungi are not plants at all. But, since they do occur in the college campus habitat, they are included in this discussion. Non-green and nonphotosynthetic, these function as recyclers, decomposing organic material for reuse.

PROCEDURE

Plant Collection

During one laboratory session, the class will take a botanical tour across the horticultural and naturally occurring flora of the CCC campus.

Designated plants will be described according to leaf, flower, fruit, or stem structure—depending upon which features happen to be present. Information regarding *ethno-botany* (cultural uses) may also be discussed. Careful observation of selected plants will be made using hand lenses.

Your instructor will clip enough samples of each plant, so that each student will make her/his own field collection. Be **conservative** in your collection: *do not take more than you need.*

The entire twig, with attached leaves, flower, or fruit will be needed for herbarium specimens.

Plant Pressing

Returning to the laboratory, the samples collected will be preserved in plant presses in the following manner, as demonstrated in class:

1. Several plants are placed within a **newspaper "folder,"** such that they are not crowded, but neither is space wasted.

    Every newspaper folder **MUST** be labeled with the student's full name.
2. 2–4 of the folders are placed in between **blotter paper.**
3. Blotter paper is placed in between **corrugated cardboard.**

    The layers are arranged as a loose alternation of:
    a. Cardboard
    b. Blotter paper
    c. Several newspaper folders
    d. Blotter paper
    e. Cardboard

4. The press is then secured with straps as tightly as possible. This is best achieved if the press is placed on the floor, and one student stands on the filled press. Then, several others can tighten the straps to the max.
5. The press should remain this way for several days. The newspaper and blotter paper will absorb excess moisture, preventing the specimens from molding.

Plant Identification

When the presses are opened several days later, the plants are to be identified according to both:
- *Genus species*
- Common name

Each lab table will have available several *guide books* for identification. Check the *index,* and *distribution maps,* in order to reach the correct identifications. The best success will be achieved if members of a lab table work together. Your instructor will also circulate and assist.

The following guide books will be available in lab for student use:

Brockman, F., 2001, *Trees of North America,* 280 pp. St. Martins Press. ISBN 1-58238-092-9. $14.95

Cobb, B., *A Field Guide to Ferns and Their Related Families: Northeastern/Central North America,* 281 pp. Houghton Miflin Pr. ISBN 0-395-97512-3. $18.00

Conard, H. S. & P. L. Redfearn, Jr., *How to Know the Mosses and Liverworts,* 302 pp. McGraw-Hill Pub. ISBN 0-697-04768-7. $38.50

Peterson, R. & M. McKenny, 1996. *A Field Guide to Wildflowers: Northeastern/North-Central North America,* 420 pp. Houghton-Miflin Pr. ISBN 0-395-91172-9. $19.00

Symonds, G. W. D., *The Shrub Identification Book,* 379 pp. Wm. Morrow Co. ISBN 0-688-05040-9. $20.00

Symonds, G. W. D., *The Tree Identification Book,* 272 pp. Wm. Morrow Co. ISBN 0-688-05039-5. $22.00

Students interested in continuing their plant identification outside of class session are encouraged to investigate the following options:

LABORATORY TECHNICIAN

Limited numbers of field guides may be borrowed from the Laboratory Technician's office, for use during the day, in the Taft building. **No overnight use.**

FIELD GUIDES ON RESERVE IN CCC LIBRARY:

Most of the field guides are on library reserve, and may be borrowed for 3-hour use within the library. Additional field guides are available from the library stacks on the third floor—Dewey Decimal Index #QK. . . . .

INTERNET WEBSITES: (keep surfing for many others)

Pine Barrens plants:

http://www.georgian.edu/pinebarrens/index.html

http://www.mikebaker.com/plants/haveflwr.html

Native New Jersey Plants:

http://www.rce.rutgers.edu/njriparianforestbuffers/nativeplants.htm

http://www.npsnj.org/photo_gallery.htm

Native Pennsylvania Plants:

http://www.paflora.org/

http://cal.vet.upenn.edu/poison/index.html

Mounting of Herbarium Specimens

Once correct identifications are made the plants are mounted/glued on **herbarium** paper. The plants should be grouped together according to the two general categories:
- ***Horticultural***
- ***Naturally occurring***

Plan ahead, laying the plants out prior to mounting. Then, be neat and careful in your application of glue.

Depending on the size of the specimen, several samples are placed together on an ***herbarium*** sheet. These should not be crowded, and space must be left for indicating:

- each plant's **scientific** and **common** names
- whether it is **horticultural** or **naturally occurring**

Writing the Laboratory Report

In addition to submitting your herbarium specimens, you will also write a lab report, summarizing the project.

All writing skills learned in *English Composition* classes must be applied:

- Paragraph format
- Noun–verb agreement
- Double-spaced typing

The following sections and **headings** will be represented, in sequential pattern:

A. **Introduction:**
   Objectives and background (see pp. 201–203)
   - Purpose of a Botanical survey
   - Plant migration:
     - Methods
     - Results
   - Natural vs. Horticultural varieties
   - Vertical Stratification
   - Variation within Kingdom Plantae

B. **Materials and methods:** (pp. 203–205)
   Thorough description of how the project was carried out.

C. **Table of plants collected,** arranged according to:
   - *Genus species* **always** in italics or underlined
   - Common Name
   - Horticultural vs. Naturally occurring

D. **Plant descriptions**—diagnostic, and other features of the specimens collected.

Submission of Botanical Project

A complete submission consists of the following two components, arranged according to the format fully described on pp. 204–205:

1. Typed lab report
2. Mounted herbarium specimens

An excellent project indicate attention to these qualities:

        professional   clear
        straightforward   clean

- Make certain that you have your name in an obvious location on all herbarium sheets and all lab report pages.
- Place these together in a **simple folder** which also indicates your name, class section, and project title.

In this way, the herbarium sheets will serve as ***voucher specimens,*** indicating the presence of particular plants at a specific time and place.

DO NOT:

Place any printed pages or herbarium specimens within plastic protector sheets, lamination, or other transparent cover.

Use a 3-ring binder, whether it is hard or soft-covered.

Submit a folder that is dirty, worn, or is printed with inappropriate graphics.

Attach extravagant adornments such as ribbons or sparkle-paints!

Biology Field Trip Safety Procedures

Safety procedures for the field trip are located in the safety policy in the front of this lab manual. Be sure to review those procedures BEFORE the day of your trip. ALL safety procedures will be strictly enforced.

NOTE THE FOLLOWING ADDITIONS:

As in the classroom, the following policies will be observed, since these activities are disturbing to others:

- No smoking
- Cell phones for emergency use only

Name _____ Section _____ Date _____

Exercise 33 Leaf and Tree Identification

For this project, you will collect samples of the leaves from the plants on the list. These plants are common forest trees and shrubs of New Jersey. Many of these leaves can be obtained right on the Blackwood campus of Camden County College. After collection, your instructor will show you how to prepare pressed samples, listing the common name and scientific name. By the midterm, you are expected to be able to identify these plants by sight and know the common name, genus, and species. Your identification skills may be tested in a lab practical.

Common Name	Scientific Name
Tuliptree	*Liriodendron tulipfera*
Sweetgum	*Liquiddambar styraciflua*
Sugar Maple	*Acer saccharum*
Pin Oak	*Quercus palustris*
Eastern White Pine	*Pinus strobus*
Pitch Pine	*Pinus rigida*
Eastern Red Cedar	*Juniperus virginiana*
Norway Spruce	*Picea abies*
Sassafras	*Sassafras albidum*
Flowering Dogwood	*Cornus florida*
White Birch	*Betula alba*
Rhododendron	*Rhododendron maximum*
American Holly	*Ilex opaca*
Scarlet Oak	*Quercus coccinea*
Mulberry, White	*Morus alba*
Sycamore	*Plantanus occidentalis*
Catalpa	*Catalpa speciosa*
Black Cherry	*Prunus serotina*
American Beech	*Fagus grandifolia*
Canadian Hemlock	*Tsuga canadensis*
Red Maple	*Acer rubrum*
American Basswood	*Tilia americana*
Weeping Willow	*Salix*
Honey Locust	*Gleditsia triacanthes*
Mountain Laurel	*Kalmia latifolia*
Magnolia	*Magnolia soulangeana*
Horse Chestnut	*Aesculus hippocastanum*
Willow Oak	*Quercus phellos*
Ginkgo Tree	*Ginkgo biloba*
Black Locust	*Robinia pseudo acacia*

Figure 33.1
Leaf Types

Name _____ Section _____ Date _____

Appendix 1   Periodic Table

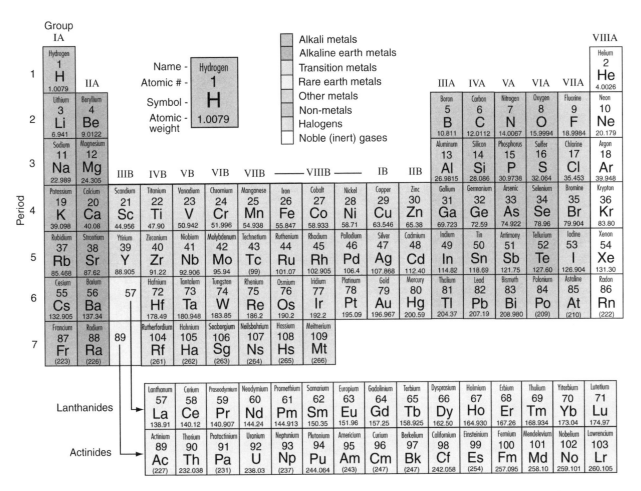

Figure A1.1
Periodic Table

© Kendall/Hunt Publishing Company

Name _____ Section _____ Date _____

Appendix 2   Laboratory Orientation

USE OF THE PIPETTE: USE A PIPETTE PUMP WITH ALL PIPETTES

The pipette is used to transfer and measure small quantities of liquid. There are two types of pipettes in general use: the volumetric pipette and the serological pipette.

    Examine the liquid in a pipette. Please note that the liquid does not appear as a straight line in the pipette. A "belly" is seen in the liquid. Water molecules are held together by **COHESIVE** forces. These forces attract similar molecules as one molecule of water to another. Water molecules are also attracted to the sides of the glass tube. These forces that attract water to the sides of the glass tube are called **ADHESIVE** forces. When the adhesive forces are greater than the cohesive forces the water molecules will tend to be pulled up the sides of a thin tube forming a belly in the liquid. This belly is called a **MENISCUS**. With most liquids we read the **BOTTOM** of the meniscus.

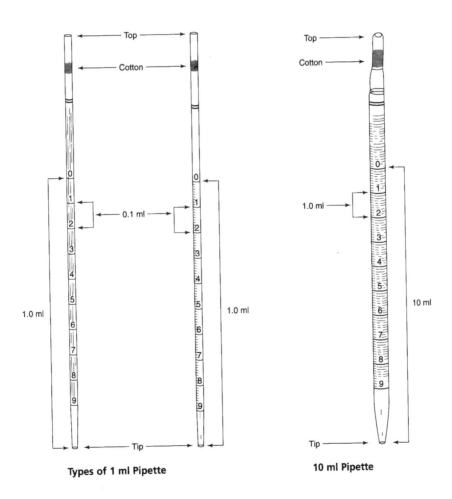

Types of 1 ml Pipette   **10 ml Pipette**

From *Principles of Biology: Biology 109–110, Seventh Edition* by R. Ragonese. Copyright 2000 by R. Ragonese. Reprinted by permission of Kendall/Hunt Publishing Company.

1. The pipette will usually come in a sealed package individually wrapped or in groups of 10 or 20. Open the package just at the very top where **"PEEL HERE TO OPEN"** is clearly written.

   If you are using the pipette in an aseptic transfer operation do **not** remove the pipette from its sterile package except **immediately prior to use.** Only handle the pipette from the **top** and do **NOT** contaminate the **tip.**

2. In this case we are not concerned with aseptic transfer so just take the pipette from the package and examine it. Unless instructed otherwise assume that we are not going to use the pipette for aseptic transfer but always open the package from the very top and never at the tip end.

3. Examine the numbers at the pipette top. You will see 10 ml in 1/10 or 1 ml in 1/10 or 1 ml in 1/100. *In all cases the FIRST NUMBER is the TOTAL CAPACITY of the pipette in milliliters. The FOLLOWING NUMBERS indicate units of gradation or the unit between TWO SUCCESSIVE marked lines on the pipette side.*

Pipette Marking	Total Volume (ml)	Volume between Two Successive Lines
10 ml in 1/10		
1 ml in 1/10		
1 ml in 1/100		

Fill in the Chart Below

4. Examine the tip. If your pipette holds 10 ml, and the number near the tip (the last number) is 9, then you have a serological pipette. If the number is 10 you have a volumetric pipette. The serological pipette holds a full capacity from zero to tip (you must release all the liquid to get full delivery of 10 ml). The volumetric pipette will hold full capacity from zero to the 10 mark. (You do not have to release all the liquid out in the tip.)

5. Place five test tubes in a rack, get a 10 ml pipette, a green pipette dispenser, and a 250 ml beaker. Fill the 250 ml beaker with 100 ml of tap water. Using the technique described, pipette the following volumes into the five test tubes.

Test Tube	Volume
1	10 ml
2	8 ml
3	6 ml
4	4 ml
5	2 ml

6. The Pipetting Technique
   a. Remove the pipette from the bag. Always leave the pipette in the bag until you are ready to use it.
   b. Remove the pipette by handling the top of the pipette only. Do not touch the pipette tip.

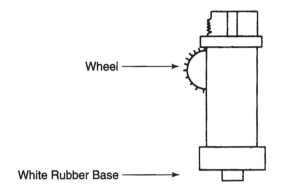

Plastic Pipette Pump

c. Place a plastic pipette pump on the end of the pipette. These plastic pipette pumps are **NOT** discarded after use, they may be reused.
d. Use the plastic pump as instructed. Place the tip of the pipette in the liquid. Turn the wheel on the dispenser until the liquid rises to the **ZERO** mark.
e. Lift the pipette from the liquid; allow the excess fluid, on the tip, to drain. Transfer the pipette to your test tube and turn the wheel to release the liquid to the amount you want into the test tube.
f. Discard the liquid remaining in the pipette into the discard container. Water or saline can be discarded in the sink.
g. Hold the pipette dispenser in one hand by the **WHITE** rubber base. With the other hand rotate the pipette gently until it is released from the dispenser. If the **WHITE** rubber base comes off, gently pull it free from the pipette and replace it back into the base of the pipette dispenser. **NEVER** throw the rubber base away. The dispenser will be useless if the base is missing.

PIPETTES THAT ARE USED TO TRANSFER BACTERIA MUST BE PLACED TIP FIRST IN DISINFECTANT (STAPHENE) AFTER USE.

QUESTION—THE USE OF PIPETTES

1. If you wished to use a 1 ml pipette to transfer 0.2 ml of fluid, what numbered line on this pipette should the liquid stop on the pipette? _____
2. If you wished to pipette 0.8 ml? _____
3. What volume is held between the tip and the 8 ml mark on a 10 ml pipette?
   _____
4. . . . between the tip and the 3 ml mark? _____
5. What is a meniscus? _____
6. How does a meniscus form in the pipette?
   _____
   _____
   _____

7. Do you read the top or the bottom of the meniscus? _____

8. Where should the pipette be kept until you are ready to use it? _____

9. What part of the pipette is never touched by hand?
   _____

10. How do you dispose of a pipette that has been used to transfer bacteria? _____

11. What is the correct way to open a bag of pipettes?
    _____

Appendix 3: Molecular Genetics

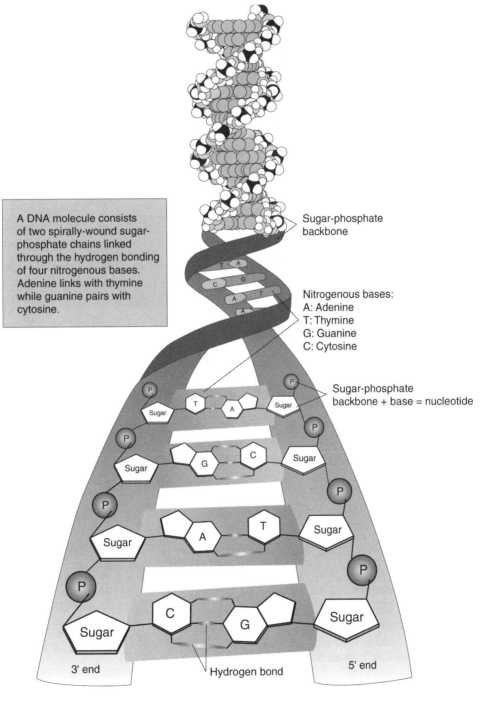

Figure A3.1
DNA Overview

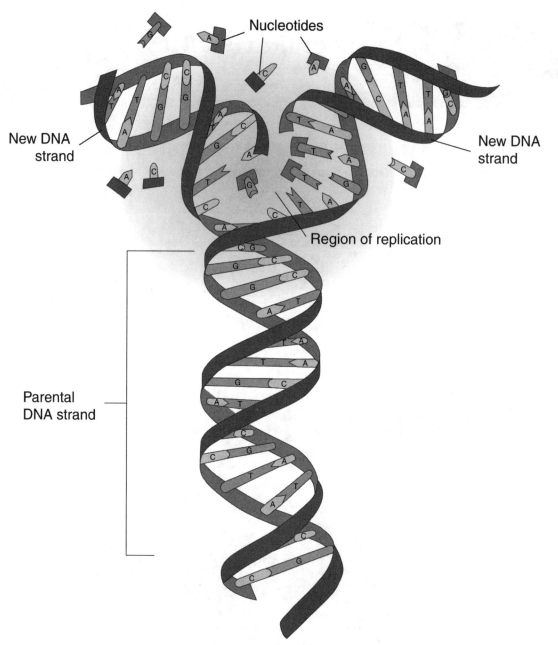

Figure A3.2
Replication

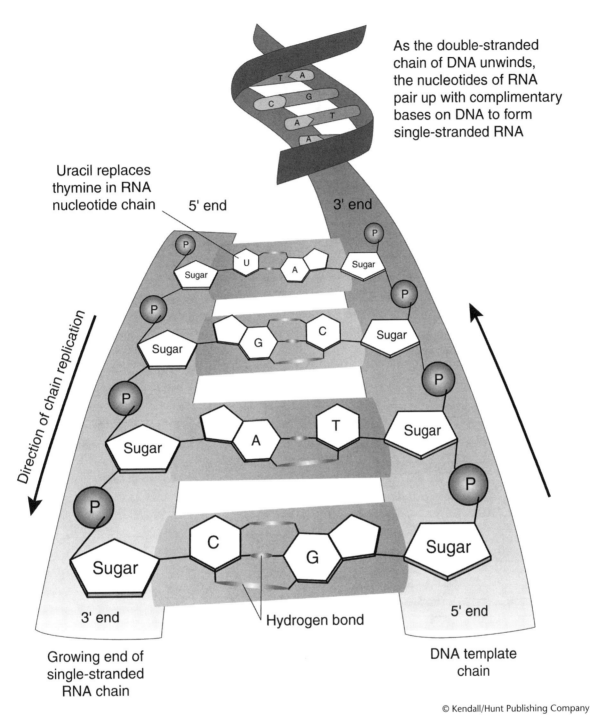

Figure A3.3
Transcription

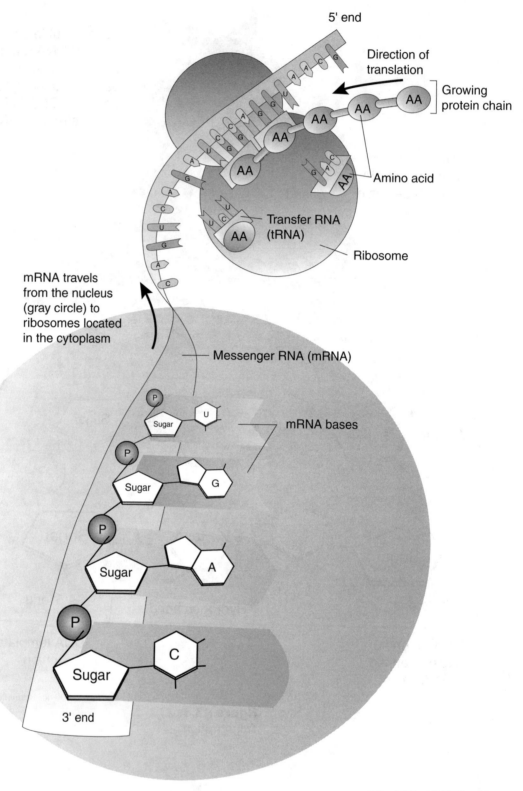

Figure A3.4
Translation

© Kendall/Hunt Publishing Company

Second Base

First Base	U	C	A	G	Third Base
U	UUU phenylalanine	UCU serine	UAU tyrosine	UGU cysteine	U
U	UUC phenylalanine	UCC serine	UAC tyrosine	UGC cysteine	C
U	UUA leucine	UCA serine	UAA stop	UGA stop	A
U	UUG leucine	UCG serine	UAG stop	UGG tryptophan	G
C	CUU leucine	CCU proline	CAU histidine	CGU arginine	U
C	CUC leucine	CCC proline	CAC histidine	CGC arginine	C
C	CUA leucine	CCA proline	CAA glutamine	CGA arginine	A
C	CUG leucine	CCG proline	CAG glutamine	CGG arginine	G
A	AUU isoleucine	ACU threonine	AAU asparagine	AGU serine	U
A	AUC isoleucine	ACC threonine	AAC asparagine	AGC serine	C
A	AUA isoleucine	ACA threonine	AAA lysine	AGA arginine	A
A	AUG(start) methionine	ACG threonine	AAG lysine	AGG arginine	G
G	GUU valine	GCU alanine	GAU aspartate	GGU glycine	U
G	GUC valine	GCC alanine	GAC aspartate	GGC glycine	C
G	GUA valine	GCA alanine	GAA glutamate	GGA glycine	A
G	GUG valine	GCG alanine	GAG glutamate	GGG glycine	G

Figure A3.5
Codon Table

© Kendall/Hunt Publishing Company